Botboy, my Botboy

Julie O'Toole

Perfectly Scientific Press
www.perfscipress.com

Copyright ©2010 by Perfectly Scientific Press
All Rights Reserved.

This is a work of fiction. Names, characters, places, and incidents either are the product of the author's imagination or are used fictitiously. Any resemblance to actual persons (or robots), living or dead, events, or locales is entirely coincidental.

ISBN: 978-1-935638-00-1

Cover art by Michael Diamond
Robot icon by Lenora O'Toole

www.perfscipress.com

Printed in the United States of America
9 8 7 6 5 4 3 2 1

Dedication

Although without the debt of love and support I owe my parents, husband, and my children I would be as nothing, this book must be dedicated to my teacher. Therefore it is for Bill E. Bond,

"Mr Bond"

1

My husband died two years ago, leaving me sad beyond telling, disoriented, and deeply at a loss. But his death also brought me an unusual gift.

Rob had been a scientist, a physicist who specialized in Artificial Intelligence, as it was called then, New Robotics as he preferred to call it. He was a quiet man, a loyal man and an eccentric man, which suited me just fine. He came from a strange, eccentric family, which suited me less fine, but as we had never had children, he cared for me tenderly and devotedly, sheltering me from his only sister and odd, odd mother, until the mother died of an overdose of weird herbal medicines cooked up in her old-fashioned kitchen. Then his sister and her equally strange husband moved a scant mile from our home in the country to raise their odd boy. They visited us occasionally, mostly to have an excuse to leave the child with someone when they went on one of their treks to far off deserts. Rob would allow the boy into his workshop, attached to the house, and I cooked for him, but mostly the child read books and kept his nose in his computer, something my husband could relate to. Rob apparently had been such a child. And then, when the boy was twelve, Rob himself died and I was alone.

Rob's gift did not come right away, though he left me a series of notes about it. He had had heart trouble and been seen by one heart specialist after another. Thus, I believe, he foresaw that he might die before me. He never spoke of it—we never spoke of it—but his notes afterward have made me believe he felt somehow that his end might be near. He perfected and programmed his final gift to suit my needs, or my needs as Rob imagined they would be once he was no longer there to fix the heat pump, to split the kindling, to fire up the tractor and plant our garden rows. It had not been completed by the time of his death and apparently it took two more years for his graduate students to get it right. He had predicted that in his notes.

As is apparently common to the recently bereaved, our big house now seemed empty, cold and hard to manage. It had been made exactly to our specifications twenty years before and had suited us very well until the hour Rob's spirit flew out the French doors of our library. His heart attack overcame him at his desk, his morning cup of coffee still steaming at his side.

It was out of the question that I leave our big country place and move to the city, for I had no family anymore and what few friends remained lived mostly in France, where I had spent my girlhood. We still exchanged recipes, in the French way, but no longer shared our days. I had been a gardener, a writer and well...Rob's wife, and now—except for the peculiar sister-in-law, her husband and teenaged son—I was alone.

When Rob first died I was primarily stunned. I felt as if my feet floated above the ground, with no anchor. When I wheeled my cart around the grocery store, I heard a distracting rushing in my ears, my vision was not quite sharp as if I could not focus on the present, but lived ever just one or two weeks in the normal past. I realized I could not buy groceries easily for I had no idea what I liked. I had for so long shopped and cooked to please my man, I had no idea what I, myself, liked to eat. And most things which I had imagined to express my long-ago French taste, were much like cardboard in my mouth.

This was the state of affairs for me for about a year, then slowly, slowly, my muse returned. I put on music, I made green tea, I wrote and read my way back to the girl who had existed before Rob, before us. I shrank a little, I dyed my hair the original color of my early girlhood, I took a horticulture class and found such consolation as the soil has to give. And then my inherited arthritis flared and I began to worry that I would not, after all, be able to stay in my home until the end of my days. Which is exactly when Rob's final, and greatest gift was delivered to my door: FedEx from his former classified lab, return address very much known.

It was a robot.

2

I didn't open the box, though I knew by now what would be inside. It was nearly as tall as I am. Rob's Notes had said that someone from the lab would be by to unpack it for me. That was just fine. Not only were my poor swollen fingers aching and too stiff to manage a pair of scissors, but, to be frank, I was filled with as much trepidation as excitement. After all, what writer needs a robot?

Late in the afternoon one of Rob's former colleagues came by, a young man named Sirhan whose eyes lit up with enthusiasm as he began to unpack and to explain. You would have thought it was a new Ferrari. He pulled down the front piece of the obviously carefully packed box and the robot slid out noiselessly. It rolled forward and stopped a few feet from my own, bleeping in a quiet way. I guess it was already activated; the young technician's eyes gleamed as he watched it.

"It rolls around the floor or any other smooth surface," he explained, "but it is perfectly capable of climbing the stairs or going anywhere you can go on rough terrain. It has a thousand little legs for that purpose underneath it... kinda like a millipede."

Ugh.

"What can it do?" I asked ungratefully, with some distaste.

He looked hurt,

"What *can't* it do? It's a social robot so advanced only NASA and... well... a few classified organizations... have one anything like it. It speaks, it vacuums, it cooks, it drives, it fixes electronic things, it... um... plays chess. I'm sure it's all there in Rob's Notes. I spent most of the past few weeks programming it to you, but it's still young."

Young?

He patted the silver side of the object and pulled a cloth hood off its "head."

By God, it had a head. Sort of. It was oval and metallic, it had wide-set compound eyes that looked for all the world like a frog or a fly, but nicer. The eyes actually crinkled up at the corners when the mouthpart curved upward in imitation of a human smile. The head was quite large. I knew from Rob that advanced robots were never made in the direct image of humans. It caused humans too much ambivalence and fright, made the robot seem uncanny to make them look like us, but robotic designers and the flock of psychologists who worked at their side knew what appealed. They created robotic heads and "faces" with the universal features of a young animal: Large, wide set eyes, a relatively big head, an engaging "smile." A robot to relate to would have to give the appearance of paying attention, of fixating on objects; it must have a "glance," be able to hold a human's gaze in order for the human to trust the interaction as a real one, Rob explained. This robot was cute, I thought, in a machine kind of way.

I hoped it would be useful, at least. And not too creepy. If it was more than I was comfortable with, I could always turn it off, or unplug it, or disconnect its batteries, or whatever you did to turn off a machine so advanced that there were only a few prototypes in the world.

"How do I turn it off?" I asked.

The young technician blanched. "Don't turn it off! This is a multimillion dollar piece of research. If you don't want it to respond to you anymore, just tell it firmly 'Sleep. Now.' And it will enter suspended animation wherever you are. But," he added hastily, "I can't really picture you wanting to do that... I mean... it's incredibly useful... fun really. Nothing you'd want to turn off. There's a manual somewhere..."

He rummaged around in the big box and came up with a thick handbook whose title page was full of equations and diagrams, nothing someone like me would consider user-friendly.

"May I come check on you and... it... next week?" Sirhan asked me anxiously. His face was red, he looked like a young man more used to talking to machines than women. I suddenly remembered Rob mentioning him, talking about his talent and enthusiasm for robots and robotic design, how he had been orphaned by the third war in the Middle East, then rescued and sent to the US by Baha'i missionaries.

"Of course," I said, as he edged towards the door, "Just call first."

The door slammed and I was alone with my glorified vacuum cleaner. I turned towards it. It didn't move, though its "eyes" followed me as I approached. Was I expected to say something?

"Hello?" I said

"Hello" came back, quietly. Not really like a machine voice at all, remarkably human. Thank God Rob had had the sense not to program his own voice into it. That would have been too much to bear.

"Welcome," I added, feeling like an idiot.

"Welcome," it repeated. "Welcome. Welcome. That is nice."

Nice? Robots knew nice?

"I understand you are a very advanced robot," I added.

There was a pause. Perhaps I had confused it.

"Advanced. Advanced," it repeated. "I am more than advanced. Advanced robots can climb stairs. I can design and perfectly execute stairs. Of wood. Of stone. Of metal. At your service."

"Cool!" A boy's voice shouted loudly behind me.

I sighed in exasperation. It was my irritating nephew and he had come in through the back door as I had forbid him doing dozens of times.

"It's a robot!" He declared with awe, "Uncle Rob told me about them and now we have one. That's so cool."

We?

Despite my frosty glare, he came closer and stuck out a hand,

"Hi! My name is Faraoun. That means Pharaoh in Egyptian. My parents named me that, but my friends call me Ron."

What friends? The boy had never had a friend in his life.

The robot put out a hand. It had five fingers which were covered with a white glove. It took the boy's hand in a gentle clasp. Apparently it did not think it odd that an annoying twenty-first century adolescent should be named "Pharaoh."

"Wow! Auntie Grace, feel his hands. They're so soft."

"I have many hands," the robot said and to demonstrate, the gloved hands were transformed into an array of shining metallic screwdrivers on one side and what looked liked pencils and drafting pens on the other. "Some are stiff, some are hard, pointed and strong. All are useful."

"What can you do?"

The robot pivoted and looked around. I was too speechless with amazement by the hand trick to stop the irritating child from persisting in his questions.

"I see that there is dust on the piano. I can dust. Or I can play the piano for you. I have been programmed to play many things." It swiveled and looked straight at me, "For example: A Foggy Day in London Town."

I felt tears catch at my throat. That had been our song when we were young, Grace and Rob.

Ron was not terribly impressed. He had spent years avoiding piano lessons.

"Can you play chess?"

"Of course."

"Can you play baseball?"

"If I have to."

"Can you clean my room?"

"I can clean any room. That is nothing."

"Yay!" Ron yelled, "A Botboy... you are a Botboy."

I frowned. I knew a little about robots from having been married to Rob. And I didn't like the term "boy," which sounded faintly colonial. "Ron. Don't shout at the robot or put ideas in its head. It is still untrained."

The robot turned a circle two or three times, looking for all the world like an overjoyed child, then it stopped to look straight at the boy.

"Botboy," it said solemnly, "I am Botboy."

"You are not." I tried to say, but Ron just danced around the poor robot and it followed him by turning noiselessly on its axis.

"I am Botboy. I am Botboy."

3

There was not a thing I could do about it. My robot was now Botboy. Not only had his name been established but his gender as well. I tried to be angry, but the more I thought about it, in those first few minutes of our life together, the more I realized that he had to be a boy, a man, a manboy. I did not need another woman in my life, I needed a man, and there he was, a priceless prototype of titanium, steel, and not a little platinum. Of course he had no real head, no torso, no legs—but what are they? He was my gift from Rob, the only man I had ever loved.

"Who made you?" I asked softly.

"Founder," he said, "Founder made me. Founder taught me about you, he programmed me to you. You are She."

"She?"

"Yes, you are called She."

The only "She" I knew, was "She who must be obeyed."

"Botboy, are you making a joke?"

He smiled. At least, his mouthparts curved upwards in what I would learn was intended as a gesture called "smile."

"Founder said: 'She.'"

"What's my name?" Ron asked.

Botboy turned back to the child and gestured with his hand as if on stage:

"Faraoun."

"Everyone calls me Ron."

"You are Faraoun."

"You see. You need to be careful at first." I told him, "Once you program a new robot with primary things, like names, they are not easily reprogrammed."

Rob had taught me that.

"I can be reprogrammed. But not by Faraoun."

Did I detect humor?

Botboy hummed a bit, a tuneless, soft humming, not unpleasant. "She. I will begin work now. You rest. I will inspect the house. There may be laundry to do."

Maybe?

"There may be dust. Perhaps the bed linens should be changed. Or the windows washed. There are only four hours until dinner, so I will need to start."

I was at a loss for words. Even Ron didn't know what to say.

"Uh...OK. You do that. I think I will go grocery shopping. I'll be back in an hour. Will you be OK without me?" I added lamely.

He turned from his inspection of the living room to look directly at me, "I will always be OK. I am programmed OK. But you will not always be OK without me. You should not go shopping without me. I will park your car. I will carry your bags. I will do a price and nutritional analysis on the food you wish to buy."

"Good grief. And what will I do?"

"You will command me."

"OK. Then I command you to stay here while I go shopping."

"As you wish, She."
Did I detect hurt?

4

I did go shopping by myself, but I have to admit that I cut the experience short to get back home. The robot seemed almost like a new person staying in our house... my house. I discovered as I drove back that I had lots of questions to ask him. The fact that Rob had programmed this robot for me made me feel eerily as if there was still something to know about my Rob from his own creation, though surely that was just wishful thinking.

I had trouble loading the groceries in the trunk of the car because my joints were so painful. Perhaps the robot—Botboy—could unload them for me? Would I tire of telling him every little thing to do?

As the car pulled into the driveway the front door opened. Botboy rolled out, easily negotiating the steps, followed by my nephew who had to lope in an ungraceful, adolescent way to keep up with him.

"Auntie Grace, it's incredible! Botboy knew you were coming; he just knew by some weird instinct that it was you before you drove up."

Botboy stopped, whirring very softly.

"Instinct? I am programmed to be flexible. I do not have instinct, that is for insects. There is nothing remarkable about knowing when She is nearing the house, even dogs can do that. My super-hearing noted the soundprint of her car as she drove away. So I heard her returning before your human ears could detect the sound."

He reached a hand for my keys. His hand, at least that iteration of it, was soft and gloved. I gave him the keys, but he held onto my hand. Very gently, very firmly.

"You have joint pain."

"I have arthritis, Botboy."

"I know this. I can help you. I have been programmed to be a healing robot. The first of its kind."

Was that pride I detected?

Ron was so excited he was nearly jumping up and down. I found, oddly, that the annoying effects of my nephew seemed not as bad, now that I had "someone" to share them with me.

"So...how does that work, Botboy?" Ron asked excitedly, "I mean can you just touch someone and they are healed? Like a medicine man?"

Botboy appeared to think for a moment, at least he did not reply at once.

"I am medicine man. Medicine man. Yes, that is true. I am a highly trained diagnostician. I carry within me all the written records available to the public domain on medicine and medications. My processors integrate them within seconds, if needed. I am not quite as fast as a fast human, but I have a larger data base. I am not as fast as Founder. But he was not a doctor. I can perform some treatments, but mostly I can direct you to evidence-based procedures to have done elsewhere."

He had not let go of my fingers. It was strange. Though he appeared to have five digits, each digit was massaging my joints as if made up of thousands of tinier digits. The effect was soothing, but I jerked my hand away. I would tell Botboy what to do, not the reverse.

"Botboy, please take the groceries inside."

Please?

He inclined his heavy silver head and moved towards the trunk. With no apparent effort he slung the cloth bags over his metallic arms and rolled back through the front door, Ron and I in his wake.

Once he had all the bags in the house he stopped moving, appearing to wait for my command. Why on earth had I bought so much food? Even after two years it was hard to shop for one.

"How would you prefer I organize the groceries, She?"

"What do you mean?" I asked.

"I can analyze your storage method and imitate it. I could alphabetize things. I could color code them."

"Well... how will you most easily find things again?"

"I will find them regardless of the method. It is a question of how you will find them."

"For God's sake, don't alphabetize them... analyze my method, I guess. But... what if I have no method? I'm not a robot. Humans can be spontaneous, you know. Not like a machine."

"Can you be spontaneous, Botboy?" Ron asked.

The robot seemed to almost sigh. At least, a small puff of air preceded his answer.

"No. I cannot be spontaneous. That is human. But humans are not nearly so spontaneous as they think. There is almost always a pattern to their actions and speech, though they are not aware of it. I can analyze their patterns, even when they cannot.

Analysis is a weak point for many humans. Robots are good at it."

"So what are humans good at?"

"Many things, Faraoun. Making robots."

"Can you teach me to make a robot, Botboy? Or better yet, just make me several for my own use?"

Botboy began to put the groceries away, not missing a beat, efficiently, if somewhat jerkily, placing everything where I certainly would have.

"Some things are forbidden. To forbid things to robots, you do not program them in. Or you program a stop. Robots may not make other robots. That would be replication. It is forbidden."

"Darn! I need a robot!"

"I think you do not, Faraoun. You need to go to school. All fourteen-year-olds need to go to school. Their knowledge is...embryonic."

Oh man, I couldn't have put it better myself.

"By the way, Ron," I asked, "Hadn't you better get on home for dinner? Won't your parents wonder where you are?"

He shrugged. "Nah. They're out of town. They went to the Mojave for a week."

I felt an old anger rising in me. They clearly expected me to be around for the boy, though they hadn't said anything.

"I'm fourteen. I can take care of myself."

"Leaving a fourteen-year-old boy alone for a week without adults is dangerous," the robot said. "But do not worry, She. I will follow him home. Once I know the way I can check on him. It can be a minor part of my duties. But only if you wish."

I did wish. What a relief.

"Can you clean my room, Botboy?"

"You can clean your own room, Faraoun. I take orders from She."

The groceries were already put away. I noticed the counters had been wiped and the floor swept while I was gone.

"I am going to take a nap," I announced, "Botboy, please follow Ron home." The life-long habit of "please" was hard to break. I refrained from asking if he could find his way back. Clearly he could.

"I will lock the door behind me, She. Whenever I am gone I will lock up, but as long as I am in the house you will always be safe. No human is as strong as a robot."

I watched from the kitchen window as the two of them crossed the yard, the boy slightly in the lead. They looked for all the world like friends.

5

For the first time in two years I slept deeply and sweetly. Rob and I had been married for 30 years. We had rarely been apart. Thirty years of hearing him breathe beside me, thirty years of hearing him putter about the house, knowing my man was with me and I was not alone. Rob, techno-lover that he was, had programmed a lot of security features into the house, but once he was gone they seemed insubstantial and unreliable and I often slept poorly, awakening at every sound. I had thought about getting a dog, but my joints were too painful to control one; a cat would clearly be no good. And now I had a robot.

I awoke briefly to hear Botboy moving about the living room, his soft whirrs giving away his presence. *As long as I am in the house you will always be safe.* And I slept again. Sweet sleep.

I awoke to the smell of cooking. I was somewhat alarmed to think what it might be. Rob had never cooked, so for him to program a robot to cook well seemed unlikely.

And yet. It smelled wonderful. Pasta Puttanesca.

Boyboy was stirring the sauce when I went into the kitchen. I knew my hair was in complete disarray and there were holes in my sweater, my eyes were puffy and the skin on my face an elderly woman's sag. But Botboy was a machine. He could not care. Notice perhaps. But not care. The perfect house guest.

"How did you know how to cook this, Botboy?"

"I have a list of 150 of your favorite dishes, She. The recipes are in the public domain. Founder thought you could alter them easily to suit your actual style once I had made them for you."

"It looks a bit much food for one."

"The boy had no food at home. I thought you might want to feed him. If not, I will save it for your lunch tomorrow."

I thought for a moment, but before I could answer he went on, "An alternative choice would be for me to take him some. He will be hungry, I estimate. Fourteen-year-old human males need about 2800 calories a day, if they are not too active. They are often hungry."

"Well, I'm hungry" I replied. "I suppose we could have Ron walk back over for dinner. You could call him? Do you know how to use a cell phone?"

Botboy looked at me, "I am a cell phone."

I hadn't thought of that. "Do you know his number?"

"No, She. Apart from any numbers you care to instruct me to acquire, I have only one programmed in now."

"What number is that?"

"911."

Ah! Of course.

"Well, let's call Ron then, so we can eat." I wrote the number on a piece of paper on the counter and Botboy picked it up.

"Do you need this piece of paper anymore, She?"

"No. Not really."

He popped it in his mouth and it disappeared.

"Good heavens, now you're a garbage can as well?"

"I am a carbon recycler, She. I can personally dispose of small amounts of carbon-based materials for recycling, such as that piece of paper. Once I have extracted the data from it, with my eyes—by reading it—the larger amounts of carbon you and I can deal with together."

"What do you mean?"

"You drive an old hybrid car, She. We will analyze all new information about cars in order to control emissions and reduce our carbon footprint. When appropriate we will buy a new car."

"But I like my car!"

"The car is just a machine. Like me. But much less valuable."

Did I detect smugness?

"Reducing our carbon footprint affects the Earth, which is a living thing. Like you."

I had to remind myself that my husband had programmed him.

"You sounded smug there for an instant, Botboy."

"Smug. Smug." He repeated. "An interesting word. From the Friesan, also Scandinavian, *smuck* or *schmuck*. Meaning self-satisfied. There are other meanings, but they are archaic. Smug cannot apply to robots, She. Smug is a feeling. Robots do not have feelings. That is human. To feel is human. To suffer is human."

I sniffed. "When I need a lesson in entomology, I'll let you know."

"Entomology is very important, She. Without it I cannot decipher human speech. Founder devoted a lot of memory space on my central processors to it. He programmed me to speak about it. He thought you were interested in it."

"I am. It's just annoying to have a machine lecture me about it."

He made no response to this.

"So...If a robot cannot decipher speech, it can't understand humans, do I have that right?"

He hummed a little more. "Not exactly. Body language and smells tell many things. Dogs, for example, know many things about humans through these clues, though their understanding of humans is limited because of their inability to decipher much speech."

Smells? Before I could delve into that one, Botboy began dialing the number I had given him. I heard a voice. It came, apparently, from deep within the robot, but it was my nephew.

"Yeah?"

No manners, that kid.

"Faraoun, Botboy here. She has requested that you come over for dinner. Now."

"Oh wow! Great! I'm starved."

No thanks, no nothing. Ron had just apparently hung up.

"He will be here in a very few minutes, perhaps eight, on his bicycle," Botboy told me.

"How do you know?"

"He is starved."

It took Ron seven minutes exactly.

See? Robots aren't perfect.

6

It took some time for me to train Botboy to do everything I wanted done, though he was a quick study, a very quick study. I needed a robot to clean the house, to cook and clean when I did not feel like it, and to stand quietly by when I did. I emphatically did not need one to talk a lot. Botboy, although capable of talking endlessly, basically only responded when engaged by me first, unless it was necessary for him to avert a catastrophe (such as the kettle boiling over) or remind me of things I needed to do.

I will walk a mile to avoid people who talk a lot. Rob had known that about me, of course; he was himself no different. Some people like the sound of a TV in the background for "company" when they are alone. I had never understood this before, until Rob was gone. Now I was alone in the morning, alone all day, alone when I awoke in the middle of the night—unless I left the house and actively sought the company of others. It was not my nature to do so, a trait that had only worsened with age, but I had reached the point where I had to have the radio on all the time to avoid the echoing, aching silence of no-Rob that now filled our house. But that was before Botboy. Before my robot came to stay.

The mere fact of his presence, the feeling of another being around, lessened my loneliness considerably, foolish though it may sound, to project "being-ness" onto a machine. After all, he moved about the house doing everything I could wish. He carried on conversations more reasonable and interesting than many people I knew. He hummed and whirred quietly, a feature I suspect Rob had programmed into him more out of a sense of humor about his being a robot than as any byproduct of his inner workings. These days many (much simpler) robots are programmed to move without making a sound, these are the ones you see at the airport or sweeping the floor in hotels. I did not need more silence; silence filled my every waking hour.

As a young writer I had been a science writer, though Rob always made enough money that I could enjoy the luxury of writing from home without the necessity for travel. I wrote about what interested me, for the New Yorker mostly and occasionally for National Geographic. Now I did none of that. I was writing a monograph on the butterflies of the Pacific Northwest and this interested me much more than reporting on other people's endeavors. I also collected, pressed and categorized wildflowers of the region; a very 18th century interest, but one that gave me great happiness. And I gardened. And I cooked and baked, though less so now. And I collected wild mushrooms and grew some at home. Rob and I had both been interested in that.

It was spring when Botboy first came to me, and that first spring gradually turned into summer. There was a lot to do outside.

Somewhere I read that Charles Darwin felt best, was happiest, when his daily routine did not vary, when he rose at the same time, ate his meals at the same times, studied and worked during predictable times only and went to bed at the same time.

He took a walk every day, rain or shine, along the same path that he had built around his property, around Down House. It led past his greenhouses, through his small woods and along the edge of the fields. Down the path and back he went each day, before lunch, before work. The blissful regularity of it soothed his mind. I wondered what Darwin would have thought of a robot like Botboy. I wondered if the great scientist would have imagined robots evolving. Did evolution require replication—that thing forbidden to robots? I would have to remember to ask Botboy.

My day, while not as regular as Darwin's, was close. I arose at 6 am, as I always had. Botboy made me coffee and the smell of it drew me from my bed as if Rob had made it: Day in, day out, year in, year out, until the days were gone... And he was gone. I read every morning and every night before bed. I had to. I had done this since my girlhood and it was as essential to me as breathing fresh air. If I didn't do it, I felt ill. Then I took a shower, put on my jeans (people of my generation never considered themselves too old for jeans) and went outside. This I did daily except in the most inclement weather. I walked around my acreage and checked on every living thing. *The best fertilizer* the Chinese say *is the gardener's own shadow.* My land was well fertilized. My shadow fell on everything.

First I inspected the plants in the front of the house: The windmill palms, the Goji bush, the Mulberry tree, the lime-green, swaying Euphorbia, the sweet Woodruff growing like a carpet of loveliness under all the trees and bushes, the rhododendrons Rob had dragged here for me from a deserted farmhouse, the azaleas in raging fuchsia bloom, and more. Then I walked to the chicken house and watched the girls clucking happily in the thin morning light. I tossed them some lettuce or beet tops, whatever I had.

One... two... six... seven. No, the coyotes had not taken any of them in the night. The hen house Rob had built for me was stout, but all things age and I worried that one day it would need reinforcing. My path took me past the hedgerow we had planted with fragrant bushes, with Eucalyptus and Willow, with Jerusalem artichokes and Seaberry. Next I opened the gate and inspected my vegetable garden, then roses, orchard trees, Hydrangea hedge, fish pond and back to the house again. A lovely, nostalgic walk.

Now I did not go alone. Botboy came with me, moving with ease across the uneven ground, quiet unless necessity demanded or I spoke to him.

"Shall we water the new plantings around the pond?" I asked.

Botboy stuck a finger into the soil.

"She, this is Rosa 'Sharifa Asma.' It is not yet dry, but it needs more sun. I can dig it a better hole over there. And it needs magnesium."

"OK."

I did not argue. It made perfect sense that Rob would have designed me a robot with encyclopedic knowledge of horticulture. My worn-out back was happy to hand him the shovel and marvel at how easy it looked when a machine moved earth.

We moved the rose. Botboy carried the pail of vegetable scraps for the compost heap; he quickly and efficiently turned the tangle of weeds and yard debris that was slowly decomposing and had not been touched since Rob was alive. I had asked my nephew to do it once, but I could not bear to beg for help.

Botboy wound up the hose I had left lying on the ground when my energy had been exhausted the day before. I followed behind him, cutting here a rose, there a phlox. The day was mild and warm.

Once the morning's work in the garden was done, we went inside. I read and Botboy did whatever tasks I set him, or stood quietly in a corner of the library and looked on, benignly, I thought. When I worked on my flower collection he rolled near to be of help. Things he did not possess in his database he researched on the web, passing along information with a comment as to how reliable he felt the web source was or was not. I surfed the web for answers myself, but not nearly as quickly.

One morning I discovered that some animal had broken into the chicken coop and taken two of my hens. I felt bereft far beyond the loss of two chickens. It was as if pieces of my life with Rob were disappearing before my eyes. We had bought those little chicks as cheeping balls of fluff and I had carried them home in a box, riding beside Rob in his old-fashioned pick-up truck, the kind they outlawed last year for its horrendous carbon footprint and usage of fossil-fuel gasoline. When I discovered the break-in Botboy was back in the house, refreshing my cup of coffee, so he did not see my tears. I dried them as swiftly as I could before I heard him coming up the path. Not that you can hide sorrow from a robot who can smell emotions, but I still have some pride and not letting a robot see me weepy and weak seemed important to me.

I showed him the spot where the animal had entered, the chicken wire had been worked back from the old wooden frame. "I don't know if it was a coyote, a fox or a weasel," I said sadly.

Botboy lifted his head as if sniffing the air. "We will find out, She. I liked those chickens."

"How can you like things, Botboy?" I asked, curious despite my sorrow, "I thought only humans have emotions."

He nodded as he bent to explore the break with one of his many-tooled hands.

"Although it is true that only humans have emotions, robots are able to divide objects into categories so well that they may seem to. For example, I have a 'good' category composed of things which benefit you. If something is 'good' by this definition, I can tell you I 'like it.' It is shorthand, in a way, and improves our HRC. Hence I 'liked' the chickens, for they are good for you: They make you happy, they eat worms and bugs, they provide fertilizer for our garden and they give you eggs."

I knew that HRC was *human-to-robot-communication* because Rob had talked about it a lot. Humans, Rob said, naturally anthropomorphize in order to rationalize behavior, to make it seem understandable. I knew we did this even with stones and trees, much less with robots who had language skills and humanoid parts. The way Botboy referred to "our" and "us" also strengthened the bond my brain naturally formed with him as "my" robot.

"Are you a computer?" I asked him once.

"I am many computers. A robot like me is complex, though not as complex as you are."

Hmmm... That sounded remarkably like male humor.

"I am just a simple woman," I said, to see what he would say.

"There are no simple women, She. There are no simple humans. Alone the hormones! Human brains and human bodies are immensely complex and interconnected. Also, humans are intricately connected to other humans, in ways not seen with robots."

"What do you mean?"

"Humans have relationships: Complicated networks between two people or more than two people. You have a mother, or did once. Founder was your husband. You liked him. Faraoun is the child of Founder's sister. You do not like her."

"How do you know?"

"I can smell it. All human emotions give off an odor. I can detect that odor in minute quantities and interpret it. It is very reliable. More reliable than speech because it cannot be controlled."

"What else do you smell?" I asked, almost afraid to ask.

"Many things. I smell...which word?...ambivalence? Confusion? When Faraoun is around you."

Hah! He got that right.

"I smell loneliness...sadness...sometimes at night when you are asleep."

"How about when I am around you?"

Now he whirred a bit.

"When I first came out of my box I smelled fear. Also excitement. But now I smell...comfort."

I sighed and reached for my cup of green tea.

"Comfort is good, Botboy."

"Yes, She, for humans, comfort is very good."

That night Botboy stood next to the coop with his motion-activated video camera on, a perfectly motionless, silent heap of metal. When he showed me the film in the morning it had captured a little red fox returning to the scene of his crime for more.

"So do you like foxes, Botboy?" I asked, to test him.

"I do," he said solemnly. "They are your favorite animal, She."

I didn't need to ask how he knew.

7

It had become usual for Faraoun to spend the evening with us (I had taken to calling him that myself, in imitation of my robot). His parents had returned from the Mojave, but even when home did not pay much attention to him. They certainly did not cook dinner, both being naturally lean and fiercely determined to stay that way. The rare times I went to their house, the refrigerator was filled with skim milk, bean sprouts, lettuce and bottles of water (outlawed in some States for their carbon footprint). The cupboards, those disordered ones which showed signs of having been ransacked by a child, were full of instant Japanese noodles and artificial flavor packages, referred to as Top Ramen.

You might wonder why, given the circumstances, I had never tried to make life easier for my nephew. Rob and I had been pregnant once and our own small boy had died in his 5th month of pregnancy, devastating me to the point where I dare not try again. I turned away from children, and the thought of children, permanently, it seemed. Rob himself was kind and companionly to Faraoun, but much too busy with his science to parent anyone properly. And anyway, why the heck didn't Faraoun's own parents do it?

Faraoun was a physically unprepossessing boy: His hair was black and flat, his skin dark, his eyes large, his frame skinny and slightly hunched forward. Pharaoh, indeed! He loped rather than ran and he fidgeted whenever still. Raised by himself he had appalling manners and was very loud. In the past whenever he was around he did that thing I dislike most: Followed me around chattering. But now he followed Botboy around, and funny thing, he talked less. Botboy advanced the theory that as Faraoun had more to say, he actually spoke less. Sounded good to me.

Once when I referred to Faraoun's rather unprepossessing appearance, Botboy replied in a serious voice:

"Who am I, a robot, to judge the appearance of a man?"

And with his large silver head, mouth that ate paper, transformative hands and no legs, I could see his point.

"But beyond that, he has no manners." I said.

"Manners are human social conventions, designed to make human interaction frictionless and to minimize physical competition and conflict. They must be taught."

"Well, perhaps you can teach him, then. I certainly won't, and his parents have failed there as well."

"Yes, Faraoun's parents have failed at many things, She. But he is not a throw-away child. He is actually very intelligent, though somewhat deficient in the social skills area. This is common among the children of scientists. Founder himself noted this about his family. And Faraoun is his family."

Put like that I almost felt a twinge of affection for the boy. You see what talking to a social robot can be like? You start by telling them to do the dishes and before you know it you are imitating their behavior and taking their advice.

"I never thought of Faraoun as particularly smart, Botboy, He gets very bad grades in school."

"He is fidgety and inattentive. He is bored. He is undernourished and has insufficient fat in his diet; this worsens his inattention. I hope you do not mind if we feed him. We always have more food than you can eat."

"As long as you are doing the work, why should I mind? I just can't bear how much—and how loudly—he talks."

Boyboy sighed.

My Rob had paid such attention to detail that he had programmed non-verbal expressions into all his robots, slight human gestures, like the drawn-out exhalation we call a sigh. It made their speech seem more normal and gave emphasis to their words, he explained. Curious how even knowing this did not remove the effect.

"She, he talks so much and so loudly for attention. No one hears him, no one pays attention. But I will do so. Then you will see that he will talk less and say more. That is... with your permission. For I am your robot, She."

"You were programmed specifically for me," I said with a little pride, a little sadness, a little wonder.

"Most of my program is generic, it would work as social robots do, for anyone, but a small yet significant overlay was programmed by Founder just for you. He felt he knew you."

Ah, yes, my Rob had known me.

It was a summer evening. The frogs that I had so carefully encouraged were singing in my rainwater-catchment pond. The air was soft. Botboy was making buttermilk biscuits and I was cutting vegetables. I had now learned that if I let Botboy do everything I would be bored. Doing things together made me much happier, especially as I could quit anytime I wished or anytime my swollen fingers made it too hard to go on. Everything would get done anyway.

The screen door slammed. Only Faraoun entered like that.

"Hey Botboy! Hi Auntie Grace. That smells good."

Well, this was a small improvement in manners, anyway. Contact with my robot was rubbing off on him.

"Where are your school books?" Botboy asked.

He shrugged. "You don't expect me to do homework over here..."

"I surely do," Botboy replied prissily. Now my robot was imitating me.

"Return at once on your bicycle and get them," he said, not missing a beat with his biscuits.

"How do you know I rode my bike?"

"I analyzed your approach, of course."

"Well..." Faraoun looked around for a distraction, "I have a flat tire."

"Really?" Botboy stopped midway between putting the biscuits in the oven and replaced them on the counter. He rolled past me and past the boy, out the door.

"Let's see."

I walked to the door, mostly to see how Botboy would handle this obvious falsehood. But the old bike, leaning against the house, did have a flat tire. Botboy "knelt" in front of it. A thin protuberance, looking for all the world like a tire gauge, extended from his hand. He began to inflate the tire.

"This will get you home and back. After dinner we will fix it."

"Nah. You fix it for me."

"I will teach you how to do it, Faraoun."

The boy shrugged, annoyed that he had to return home for his books, but unable to resist dinner and—I was guessing—his time with the robot.

"OK. Do that then, Botboy."

Botboy turned and looked squarely at the child with his large eyes.

"I may be a robot, Faraoun, but you will say 'please' to me. If you will say please to a robot you will remember to say please to a human."

"And if I don't?"

"Then I will not do what you ask."

Faraoun shrugged, "You only do what I ask if *She* says so anyway."

"That is true," Botboy said and was silent.

Faraoun shrugged again and went outside to the bike. Despite his reluctance he pedaled furiously and returned as the biscuits were just out of the oven, smelling, I must say, like a small piece of heaven. The stew was ready, we stirred in the last-minute vegetables.

"Yum!" Faraoun declared.

Botboy did not turn around, "Wash your hands and set the table, please, Faraoun."

Faraoun stared. "Wash my hands?"

"Do you need a lecture in microbiology before you eat?" Botboy replied.

Faraoun washed his hands. And he set the table, something I had never seen him do before. He was at a complete loss to know where the spoons and forks went, so I showed him.

"You mean there's a right way to do this?" He asked, astonished. "Why does it matter?"

"There's a conventional way, surely." I replied. The poor child so rarely ate at a table with adults.

After dinner, Faraoun actually helped me clear while Botboy did the dishes. Then Botboy herded him to the table with his books. Faraoun began to whine.

"It's boring, Botboy. Let's play chess instead. I can get smart playing chess."

"Faraoun. You do not know enough to decide. That is where adults—and robots like me—come in. Your time on earth is much too short for you to spend it resisting your greatest human gift."

"Oh yeah? What's that?"

"The ability to learn. Faraoun, did you admire your uncle? Founder?"

"More than anyone," the boy said seriously. I felt another unexpected twinge of affection for him.

"Founder had two PhD's, Faraoun. He taught many students, he built many things. He built me. You are his close genetic relative. You can be like him. Humans are mostly like their closest genetic relatives."

For better or for worse, I said under my breath, thinking of Faraoun's parents.

"If you learn things you could be a master robotitian. Or even a robotologist."

"What kind of things?"

"Math to begin with, and physics and biology. The most advanced robots are social robots. For that you need to understand biology, especially neurobiology. The brain. It is all about the brain."

"Well why can't you just download all this stuff directly into my brain for me?"

Botboy paused. "That cannot be done yet, Faraoun. Unlike me, you are not a machine. And I think that if it could be done it would likely do irreparable damage to your neurons—your brain cells—and disrupt their other functions. Learning is the chief job of your brain. To date there are no short-cuts."

"So what is learning?"

"Learning is a process of rearranging the chemicals inside your brain into new patterns. When you take on new information and attempt to store it, to remember it, the connections between brain cells are strengthened or new connections are made. This is called learning. It is affected by genetics, by diet, by sleep and by context."

"Did you know this stuff, Auntie Grace?" Faraoun asked me, hoping, I suppose, to find an adult without the troublesome PhD's who might disagree with the robot about the importance of learning.

"Of course. I was married to your uncle for a long time. I was his sounding board."

Botboy looked up from his perusal of Faraoun's school books. "Sounding board. Sounding board. I will be your sounding board, Faraoun. If you learn."

"Do I have to get good grades?"

"I do not care if you get good grades, Faraoun," Botboy said. "But *you* should. There are only a few places in the world where advanced robotology is studied. They only take students with good grades at those places. So you would be left out. Founder would be disappointed in us both. What a pity."

Faraoun reached reluctantly for his chemistry book. Botboy put a soft gloved hand over his, on top of fat blue book. What a picture it made: The reluctant adolescent and the gentle robot.

"Start here Faraoun, I will help you."

8

How was I to guess that owning a robot as advanced as Botboy would plunge me into controversy? I hate controversy. I avoid argumentative people. I spurn politics. Perhaps I am wrong to do so, after all, I have taken advantage of an expensive Western education. Did that privilege imply that I owe society my psychic energy, an interest in politics, the social contract? At my age, the mere thought of controversy made me want to take a nap.

One day a letter came in the mail. The envelope was official and embossed. The return address was an office in Washington D.C., from the Department of Homeland Security, of all things. The name of the sender was—get this—Homer Chukkerpuppy. I thought it was a joke.

Oh it was no joke.

Mr Chukkerpuppy felt that a private citizen should not be allowed to own a robot as advanced as Botboy. He felt it was a crime that Rob had made him for me. My robot was too valuable, he said, to be a household slave.

Slave? Botboy and I had a good laugh about that one later.

Mr Chukkerpuppy encouraged me in the strongest terms to voluntarily donate the robot to the Government Agency he represented, otherwise he threatened legal action to impel me to do so. As a private citizen, he wrote, one without any security clearance whatsoever, I had no business owning a robot so advanced that certain scientists would be very interested in reverse-engineering it. *Foreign scientists* he penned ominously. I could expect further contact from his agency if I did not agree to meet Mr Chukkerpuppy soonest. Alone the robot's bomb sniffing capability and explosion-resistance made Botboy too special for a private citizen, he ended.

"What bomb sniffing capability?" I asked Botboy, who was folding laundry into precise rectangles and methodically stacking it on the flat warming plate of my Aga stove. "What explosion resistance?"

Botboy took the letter from my hands and read it swiftly.

"Ah, Mr Chukkerpuppy. Founder warned me about him. He is a robotophobe. He wants to outlaw advanced robots, not reverse-engineer them. He is notorious in Founder's lab. They hate him."

"And the bomb sniffing?" I insisted.

Botboy nodded his head as if agreeing with an unknown audience, "I am better than a dog or any living animal at sniffing out explosives. I can analyze scents and not only discern the presence of bomb-making materials, but analyze their composition. Within seconds. I can then transmit this information immediately to whatever authorities *you* command me to."

"Me?" I squeaked.

He ignored me, "That is an advanced capability, of course, but I am hardly unique in that regard. The TSA has several robots as good at this as I am, as that is all they can do. Interestingly, the technology came out of the perfume industry."

"Never mind that," I said crossly, "This man is threatening me...us. What does he mean explosive-resistance?"

Botboy's compound eyes gleamed, "Ah, She. That is something else. You know the royalty checks you get from Founder's estate twice a year?"

"Yes, of course."

"That is for merely a few of Founder's original innovations in explosive-resistance. The most important he shared with no one because those have made me what I am...meaning..." he paused dramatically, "I cannot be blown up."

"What?! Anything can be blown up, surely."

"Almost anything."

"You can't be blown up?"

"I am very...sturdy. I cannot be disassembled by any explosion short of an atomic blast."

"And Mr Chukkerpuppy thinks that is bad?"

"Mr Chukkerpuppy, and the people he works for and with, only guess that Founder was successful in creating such an entirely explosive-resistant robot. Founder's work was highly classified and those within our government whom Founder meant to know about his inventions, do know. Mr Chukkerpuppy is not one of them and they guard their specialty secrets well. Those whom Founder trusted and worked with in the government have their own prototypes of me...or perhaps not quite."

"You're very valuable then, Botboy," I said slowly.

"I am valuable beyond calculation in some ways, She. Yet Founder felt that my greatest value lay in what I could do for you. And those like you."

"Like me?"

"Older people She, those who are alone. Those who need physical help and who need companionship. He created me to fulfill this, which he believed to be the greatest gift to humanity he could offer. Far greater than explosive-resistance, useful though that may be in some quarters."

"But no one knows about you. You just live with me."

"Perhaps Founder meant you to tell them."

"Me? I'm not a robotologist."

"That is precisely the point. When people see what I can mean to you, they can imagine what a social robot could mean to them. That is what Mr Chukkerpuppy really fears. My bomb sniffing and other talents are a distraction from his real theme, which is that robots are dangerous to humans and will one day take over the Earth from them, enslaving them and making them redundant. That nothing could be farther from the truth does not stop him. But that, She, is the real controversy."

I folded the letter up. I had a notion that I should save it and not let Botboy chew it up. Could it really be that Rob had intended me to introduce robots like my own to the wider world? How could I do that? I certainly did not want to give speeches or promote either of us in any way. A web page or a blog would require hours of responding to perfect strangers online. I couldn't imagine that would be pleasant. Rob had never spoken to me about any such mission, but perhaps he had hoped to live much longer, too. Had Rob lived, he could have spread the word about social robots himself.

"Maybe his graduate students will be the ones to introduce advanced social robotics to the world," I suggested.

Botboy sped ahead of me on his return trip to the laundry room, leaving a tidy stack of tea cloths steaming on the Aga. I followed, the offending letter in hand. He put a load of towels in the washing machine before he answered, measuring the biodegradable, plant-and animal-safe soap I always used into aliquots accurate to a fraction of a gram. He pressed the "water saver" feature; Botboy was nothing if not eco-conscious.

"Think about it, She. Where are they? Why do they not come to you and ask for your—and my—guidance?"

"I can't imagine."

"Because they have come under pressure themselves since Founder left us. People like Mr Chukkerpuppy are laughed at in the scientific community, but they are taken very seriously in our governing bodies. Panic and fear about robots has been easy to spread among lawmakers and financing powers. It has become increasingly difficult for graduate students of robotology to get funding if they are working on anything much more sophisticated than a highly efficient vacuum cleaner or a faster and more accurate machine manufacturer. In significant ways, our country now lags behind the Asian and African countries when it comes to techno-enthusiasm and techno-imagination."

"But Rob could not seriously expect me to fight this battle for him? I'm not even a scientist."

"That is precisely the point, She. You are a regular woman who loved a man who died before her, whose health has been weakened, who is alone. Only someone like you could make people feel safe about robots."

I narrowed my eyes at my robot. Could he have been so well programmed as to easily manipulate me?

"Are you trying to manipulate my feelings, Botboy?" I asked. He smiled.

"Not me, She. Founder."

9

I decided to start small. There was no way I was going to give a big public speech and since my Church boasted only services of the old fashioned Roman Catholic kind, where quiet and privacy were the rule, I decided to introduce Botboy to my Horticulture Club. In general, I thought, plant people were open-minded and kind, and I would not be afraid to speak to them. This month we were going to have a social event where the challenge was to bring your all-time favorite garden tool. I would bring Botboy. As a garden tool he was not so far-fetched and this would allow me to introduce the concept of a social robot to regular people and gauge the reaction. I had the feeling Rob would approve.

I knew that the hostess for this gathering had a robotic floor duster and car washer, of which she was very proud. Because of this I felt that her house was an auspicious place to hold Botboy's first public appearance. I let her know that the "tool" I would be bringing was a robot. She was delighted.

"Ooo... if it really works and isn't too expensive, maybe I can get one too," she said over the phone. I made polite noises and rang off as quickly as I could.

Faraoun wanted to come as well, but Botboy forbade it. He was behind in his homework and Botboy wanted his Physics done by the time we came home. He told the boy he would bake him brownies if the work met his approval, and we left Faraoun sitting at the table, an old yellow pencil behind his ear, computer closed beside him.

Botboy drove. He always drove. I had joked with Rob that if I were to have only one kind of servant it would have been a chauffeur. Robotic chauffeurs had been introduced a year before Rob died, but they were just pieces of the car. The car drove itself and took primitive voice commands, though even at that early stage robotic "drivers" had proven safer than human drivers. Rob had predicted that such conveniences would spread like wildfire and in the past two years I had noticed more and more people reading a newspaper, working on their laptop or doing their hair while the car drove them to work. Nice. But Botboy as a driver was much, much nicer. Like an old-fashioned chauffeur he could speak to you, if required, he could park or hover and wait, he could locate stores and addresses and he was adept—well beyond adept—at preventing collisions. Furthermore he analyzed the engine as we drove, largely to save fuel, about which he was a fanatic.

Botboy followed me up the walkway to the front door and our hostess let us in. Nearly everyone had arrived before us, and I had planned it that way. The day was warm and people were drinking cold beers and cool white wine and eating from a lavishly laid salad bar, the contents of which, we had been told, would come entirely from the hostess' garden.

Mrs Elerina Gates was a tall woman with swept-back elegant silver hair and large pearl earrings. She dressed rather like a country squire, despite the gender implication: Riding breeches (she kept horses), boots, a simple crew neck silk blouse and a lightweight greenish tweed jacket, even in the heat. She gasped when she saw Botboy, nearly as tall as I, standing politely to one side.

"Yikes! Is *is* a robot." She screeched, "A science fiction-like robot, Grace! Where in the name of God did you get him?"

Botboy and I sailed past her, both of us smiling our most social smile. I had the feeling his was more real.

"My late husband was a robotologist, Elerina. He built him."

People had heard Elerina screech and came to see what the fuss was about.

"Wow!" "Good grief" "Look at that!" "Can he talk?" A babel of comments and questions assailed us.

"I can speak." Botboy said, using his metallic, machine-like imitation voice, not the natural one he used at home. This more primitive voice was the one he always used around strangers, he claimed it put them at their ease. People expected a robot to be a machine and were unnerved if one proved much, much more sophisticated than they imagined. A machine voice reinforced his non-human status.

"What is your name?" Asked a young woman slowly, speaking as if to a phone robot. She was the only person under forty in the room.

"My name is Botboy."

"What can he do?" Elerina asked me. "Can he help you in the garden?"

I smiled and made to get a glass of wine, as casually as I could muster. *What couldn't he do?* But I would start with what they were interested in.

I got my glass and turned to speak. Every person in the room had followed us and no one took their eyes off of Botboy. It was a bit disconcerting.

I closed my eyes and tried to channel Rob, dear Rob. "Botboy is what is called a social robot," I began. "That means he is engineered to interact with humans on a social level. He can speak, understand commands, carry out sophisticated tasks and interpret human behavior and responses."

Botboy looked at me reproachfully for this vast understatement of his capabilities and for the word "engineered." But Rob had told me that people, especially those who were uncomfortable with robots, liked to think of them as having been engineered. Like a rocketship. Sophisticated, maybe, but still a machine.

"He can weed and water and carry things. My husband programmed him with a vast horticultural knowledge."

"Oh yeah?" A thin woman in her fifties pushed herself forward. I had seen her at other gatherings, her sharp features bringing the word "scrawny" to mind, her long shapeless dress and brown clogs marking her as a latter member of Rob's and my generation. She fixed her glance on Botboy, "Then which Trachycarpus species is most hardy in zone 7 and below?"

A fat man with a vast beer belly next to her laughed and said: "Is that supposed to be the equivalent of the 'inside fly rule' for a horticultural robot?" The man had intelligent eyes and large hands, hands that loved flowers.

Botboy answered calmly, mechanically, "Many people think it is *foruneii*, but it is actually *wagnerianus*."

Several people clapped. Botboy bent at the waist in a small bow. I didn't know he had it in him.

"So he's an expensive step-and-fetch it?" The scrawny one asked, not acknowledging his correct answer and coming up close to his face. She touched his compound eye with her long finger. Everybody flinched. How often do you poke a perfect stranger in the eye, even though he's just a robot? But Botboy just blinked. I didn't worry. Anyone who can't be blown up surely can't have his eye poked out.

"You are allergic to the aspartame in your diet coke," he said politely to her after a second, "Do not drink it, it is bad for you."

She sneered, she actually sneered. I was astonished that someone would be so upset, so hostile to a robot after almost no interaction with it. But Rob had warned me about this.

"Who says I'm allergic?" She asked tightly, "What is it supposed to be bad for?"

"The disposition?" The fat man suggested sweetly.

"It is bad for the immune system, in general," Botboy told her, "but in your case, you are allergic. You don't get hives, but it will give you chronic headaches. Perhaps you have noticed them and thought they were migraines?"

"I don't take advice from robots. In fact, I don't believe in them at all." She replied, aware that we had all heard her complain at previous meetings about the migraines for which she had to take ever stronger drugs.

Elerina stepped forward, the perfect hostess, "Let's all sit down, shall we? I think the 'tool' Grace has brought really rather outdoes anything the rest of us have. But I am curious to know more, aren't you all?"

Everyone allowed themselves to be herded into the spotless living room where chairs had been arranged in a circle. Balancing my glass of white wine I sat on one of them, Botboy standing at my side. He took my purse from me, slung it over a hook on his shoulder and held my glass while I got settled. This was his Jeeves routine, and it amused him no end. Sometimes he even used his perfect imitation of a Jeevesean British butler's voice, something I had anticipated and strictly forbidden. He was disappointed, for he and Faraoun found it endlessly entertaining.

"A social robot is something new," I began. "My husband and other robotologists have been working on them for years, trying to perfect their usefulness to ordinary people."

"What do you mean by ordinary people?" Someone asked.

"I mean as opposed to, say, NASA or a large manufacturing company."

"I have a robotic car washer," Elerina added. "I think it's so cute. Sometimes I talk to it and I have given it a name. My whole family calls it that name."

"You see," I said, "The human mind is made to create relationships. It automatically endows even primitive creations—like your robot, which cannot speak or interact except in the human imagination—with powers of personality. Think of a little girl and her doll."

I went on: "Your robot is a great thing, Elerina. But a truly useful social robot should be able to do things like wash your car and also to socialize with humans, to act in such a way that humans accept them as companions and that they *are* companions. Like a dog, only much, much smarter and more human."

"Now that's where I draw the line," the scrawny one said. "Creating something to imitate a person is playing God. And I am a Christian. It should be against the Law. It is certainly against God's Law."

"But think of how wonderful it is to have help from a robot who can interact with you," I said. "If you are lonely, as... as I have been since Rob died, you have someone to be with you."

"That's what the TV is for," she said firmly. "At least it doesn't pretend to be a human. God made man in his own image. I do not think he meant for that act to be imitated by man."

"It would be hard to tell," the fat man said.

"To tell if it was like a man?" Elerina asked, confused.

"No, to tell what God meant or did not mean."

"It's in the Bible." The scrawny woman said, "Thou shalt not make graven images."

"I agree with you," Botboy said, "I do not think robots should be worshipped."

"They shouldn't even exist as far as I am concerned. If they do everything for us then we will become lazy and shiftless. If we allow them to take over our entertainment, pretty soon we won't be able to entertain ourselves."

"Like TV?" I asked. I couldn't resist.

"No," she said, "No. Like those horrible video games young boys are always playing: Violent, anti-religious and sexual."

"I do not approve of video games," Botboy said, to my astonishment, "Young boys—and girls—should use advanced technology to make the accumulated wisdom of the human race intelligible and—yes—fun."

"Why necessarily fun?" Elerina asked.

"Because learning is reinforced by context. That is why those video games are so powerful: They are fun."

"Sex is fun," the fat man added hopefully.

"So humans report," Botboy responded in his most machine-like voice. I could see now why he did it. How wise Rob had been—or maybe it was the cloud of psychologists who had always followed him around.

"Robots are the devil's invention. I don't care if they *can* weed." Scrawny added darkly.

"Botboy is a companion to me," I said, "I have severe arthritis. I probably couldn't even stay in my own home much less garden without him. I think he is a gift from God."

The younger woman stood up and approached Botboy, she smiled and he smiled back. Social Robot 101.

"My grandmother is 90 years old," she said softly, "My mother is 70 and can hardly take care of her. She has to lift her and bathe her and give her her shots for diabetes. She is blind. With a robot like you, my grandmother could stay all of her days with her family."

"That is correct," Botboy said, "and if she lives another 10 years she may be able to take advantage of Body version 2.0 and get well herself."

I groaned. Oh no. Not Body version 2.0. They would never go for that.

"What is that?" Scrawny asked suspiciously. I noticed she had put down her diet Coke.

"It is a prediction, pieces of which are already in place, where swarms of nanobots, miniscule robots, can enter the bloodstream and repair the human body from the inside: No surgery, no replacement parts, cancer surveillance which would make cancer unknown; heart repair which would make a heart attack a thing of the last century. Like polio."

Botboy squeezed my hand. We were both thinking the same thing, I know we were. Body version 2.0 had come too late for Rob, even though he had known all about it and the promise it held.

The thin woman let out a little scream: "I knew it! Listen to the Robot! He is the Trojan horse concealing swarms of microscopic bots which will take over our bodies. And the body is the temple of the Spirit! They will take over our minds, they will destroy the religion in our human hearts. We will become machines... like them."

"Yes, the body is the Temple of the human spirit." Botboy said slowly, "That is why the nanobots will let you live forever... or nearly. Like the ancient people in the Bible. Like your Methuselah."

"I know it's wrong," the woman said, sensing a trap. "It can't be Christian."

"It's neither Christian nor not-Christian," I said. "Social robots are about people and the relationships people build. If I had been able to choose I would not have needed a social robot, I would have preferred that my own husband Rob were alive."

"Me, too," Botboy whispered in his normal voice, "Me, too."

10

I should never have been on the ladder in the first place. Botboy was inside making beds (precise hospital corners a German nursing matron would have been proud of) and I was taking advantage of a glorious summer day to deadhead my largest rhododendrons.

These rhododendron bushes were more than bushes to me, Rob and I having dragged them to our property 20 years before. They had been consigned to the rubbish heap by a landscaping firm that was "modernizing" a large estate. Once they were in the ground Rob had ignored them, immersed as he was in his robotic dreams and schemes, but I babied them, watered them, adjusted the pH of their soil and stared at them every day as they slowly took hold of the earth. For the past several years I had not been as attentive though, for with the loss of Rob my energy had been unfamiliarly low and I had become afraid to undertake many physical tasks.

Since Botboy had come into my life now I could feel the welling-up of some of my old energy and some of the everyday fearlessness returned. I was no longer afraid to go for a walk in the woods, to drive my small tractor. I simply felt less "lost" than before. And the increased physical movement seemed to help my arthritis.

I reached for a dried blossom remnant and reached a little too far. I could feel the soft ground shift with the sudden redistribution of my weight and the ladder begin to go over, in slow motion, it seemed. Thank God I had the sense to cry out, for who knows how long it would have been before Botboy had come to check on me or ask me a question. Faraoun lived too far away to hear me yell anything and was in any event at school.

I felt a searing pain which seemed to come from my left arm as I landed, but I must have lost consciousness for at least a moment, for the next thing I was aware of was the presence of Botboy and the sound of his voice. He had apparently dialed 911, I heard him give our address and a brief explanation of my injuries that would have done a doctor or a nurse proud, only calmer. He had not lifted me up, but rather placed something soft under my head and covered me with a blanket. God knows where he found it.

"What happened?" I moaned. I seemed unable to move my left arm and there was the unpleasant sensation of dirt and grit in my hair and along the side of my face. I could smell the loamy earth I had worked so hard to develop.

"You fell off the ladder, She. Please do not move."

"It hurts," I whimpered.

"I will get my first aid kit from the garage."

Botboy hurried off. I could see him from the corner of my eye.

Botboy hurrying was an amazing sight, when he moved at top speed as he was doing now, his many millipedenous legs gave the impression that he floated above the gravel driveway. I heard the garage door open. He had long ago programmed the door opener to his body. It made him appear like a magician.

The pain came over me in waves like nausea, in fact, I was afraid I would vomit, so intense was it.

In a moment he was beside me again "kneeling" in the soft earth, the first aid kit open.

I slewed my eyes over to him without turning my head and watched him draw out clear fluid into one of the needle-like appendages that were a part of his "hand" repertoire. I barely felt the shot, but within seconds a sensation of floating came over me, a great indifference even to my own predicament.

"What was that?" I asked, my words slurring uncomfortably. I could hear the faint sounds of a siren on the air.

"Morphine sulfate," he answered, taking my pulse. "An ancient drug, but useful. Good shelf-life, too."

In my dream-like state I was contemplating the meaning of the word "shelf-life" and how it might or might not refer to humans. Morphine, I told myself, was good. Very good. I remembered how Botboy had told me that he contained within his computers access to a huge medical database. Another good thing. I suppose anyone could have had access to such a database via their own computer, but how would one access it lying on the ground? Interpret it once you got access to it? Such were my thoughts as I lay there, once the worst of the pain had subsided. What if Botboy had not been here, I asked myself. What if, as in the years before he came, I had been alone? Did older people die on the floors of their bathrooms, alone, cold and afraid, having slipped in a puddle of water? Or on the ground outside, having fallen as I had?

The crunch of gravel and the sound of doors slamming stopped my thoughts in this direction. The ambulance had arrived.

"What the—" I heard a man's voice. He had apparently caught sight of Botboy.

"Shit. It's a feaking robot!" Another voice.

Botboy stood perfectly still, saying nothing as they approached with their bags and stretcher.

Then he slowly moved to one side to allow them near to me, nothing too fast. When they came into my line of vision I noticed they were having a hard time focusing on me in their fascination for the robot. They seemed almost a little bit afraid.

"She appears to have broken her left humerus," Botboy said, in his machine-like voice.

They ignored him.

"Lady," one of them said, "Can you hear me?"

"Perfectly," I managed, though somewhat softly.

"It looks like you fell off the ladder." *Did he imagine I didn't know that?* He then began to speak in a professionally reassuring tone. "We are going to have to move you, but before we do that I want to start an IV to give you something for pain. Then we'll put you on the stretcher and take you to Holy Cross Hospital. OK?"

Before I could reply Botboy said: "I gave her a dose of morphine. And her insurance does not cover Holy Cross Hospital, it will have to be Deaconess Hospital. They are only 0.5 miles apart."

"What the hell..." the first voice repeated, considerably irritated. "I'm taking orders from robots now?"

I could feel the other medic tie off my right arm with a tourniquet and begin searching for a vein. "Dude," he said, "you take orders from robots every day."

"Are you referring to my wife?"

"No. Idiot. Your car. And our medication-dispensing machine. And the 911 robot who gave us directions here."

"Hummph." Metallic clicking indicated he was opening a light-weight stretcher.

My hearing was getting a bit more distant. Shock perhaps. I did hear one of them asking Botboy how much morphine he had given and his answer.

"Please," I said, "Please let my robot ride in the ambulance with me..."

"Sorry lady," the first voice said, "family members only."

Family members only. There was a concept. Botboy was actually my only family, unless you counted my nearly worthless nephew nowhere to be found. Perhaps one day when social robots were everywhere people would accept the usefulness of having them with their owners in virtually all circumstances.

"Botboy," I said faintly.

"Yes, She."

"Drive to Deaconess Hospital. Meet me there. Bring my purse. It has my insurance cards. And my extra glasses." My glasses had fallen from my face. In fact, I had heard a glasses-like crunch as I landed. Pieces of them were likely embedded in the arm where I now had little feeling.

"Up she goes," a voice said. In my dream-like state I had entirely missed that I had been placed on the stretcher, but I did notice it when they lifted me off the ground. A massive wave of pain hit me, blasting all other considerations from my mind and bringing tears to my eyes. "More meds," voice number two said.

I desperately wanted the familiar comfort of Botboy's voice, but heard nothing. And the next time I was aware of my surroundings I was in the emergency room on a hard, hard table, two nurses and a young doctor around me, apparently removing a small portable X-ray machine.

"Who found her?" One of the nurses asked.

"The field sheet says her robot," said the other.

The young doctor, a man, had been bending over me looking at my arm, but his head shot up at this comment: "Her what?"

"Her robot. And doctor. Your language. The patient is awake. This is not the operating room."

I kept my eyes closed for the comfort it gave me. Hospitals were not comforting places. I heard a commotion in the hall, voices raised in amazement, irritation, wonder. Botboy must have arrived.

"It claims to be her robot. It certainly looks like one. But it talks. Can you ask her? I mean, we can't just let some mobile machine in a patient's room just because it claims to be hers."

"He is my robot," I said as loudly as I could, though it came out as a whisper.

"Let him in!" I tried again.

The doctor had left the room, suddenly more interested in my robot than in me.

"What did you say?" The nurse asked me.

"That is my robot. Please let him in. I need him. He will be no trouble. He can understand commands."

They conferred outside my door. I could understand that they were at a loss what to do with this novelty. Rob had told me many times that early social robots, no matter how successful, would likely be met with awe, amazement and a sense of "other." People, he said, humans, do not do well with a sense of other. Wars were fought with those who were "other." Politicians intentionally created a will to destroy by identifying the enemy as "other."

"Please," I kept whispering. "Please. Please."

Suddenly Botboy was beside me. He glided smoothly across the floor and stationed himself immobily at my head. He was silent. He was taking the measure of the humans in the room and gauging how best to reassure them. He took my hand in one of his soft gloved ones. I feel asleep nearly instantly. The last words I heard were those of the surprised young doctor "Look! Her heart rate has slowed and her blood pressure is lower when that robot holds her hand."

Botboy told me later that I had been taken to the operating room to have my arm set and pieces of my glasses removed from my skin. They also did something he called "debridement" which I gather meant cleaning a wound of dirt and torn flesh. Yuck.

When they took me back to my hospital room where I was to spend one night we became something of a sensation, a constant parade of hospital personnel came by to see "the amazing robot."

The young doctor stayed the longest, apparently he was the surgeon who had set my arm. Good Lord, he looked nineteen, though I imagine he wasn't.

"We have robots in the operating room," he told me as Botboy held a glass of tepid, tasteless juice up for me to drink through a straw. "They help us with microsurgery, but they are nothing like your robot."

"Yes," I said, "Botboy is a different kind of robot. He was engineered by my husband Rob, a well-known robotologist."

I almost said *created* but it had a Frankenstein-like ring I wanted to avoid. "He is what is called a social robot. He interacts with humans in a social way, he understands social cues and speaks and gestures with near-human facility."

"And," the doctor said admiringly, "he has an encyclopedic knowledge of medicine."

"That is nothing," said Botboy modestly, "That is something any advanced computer could manage. More engineering and thought goes into a robot, any robot, being able to distinguish between the pronouns 'I' and 'you' than into any medical database compilation no matter how sophisticated or up-to-date."

"Really? Why is that?"

Botboy had dropped his mechanical voice once we were alone with the doctor and spoke in his normal, if somewhat erudite, way.

"Because those linguistic conventions imply a sense of self, of consciousness. Previously only humans and a very few non-human primates were thought to possess this. Conscious awareness implies questions about 'self' and 'other,' about singularity and commonality, about what it means to be 'alive'...what it means to be human."

Botboy must have felt this young doctor not only capable of understanding what he meant, but also non-hostile to such deliberations, deliberations that bordered on the "uncanny" for many people. Rob always said that the strongest opposition to robots would come from humans for whom a robot's very existence would open up deepest insecurities about who we are and where we come from. *How we came from* he often added mysteriously and ungrammatically.

"In fact," Botboy went on, "even the mechanics of making gestures that humans can accept as a natural part of speech is much more complicated than any medical data base."

"Say," the young doctor said suddenly, apparently tired of this line of discussion. "Can you play chess? I have a set in my locker and I'm quite good at it. Wanna play?"

Botboy seemed to consider his answer.

"Why would you want to play chess with a robot? Already in 1996 the Grand Master human chess player was beaten by a mere single-computer."

"I'm pretty good," the doctor laughed.

Botboy appeared to be considering it further. "Perhaps I could set my chess player to 'elementary,'" he said kindly.

The doctor seemed a little miffed. "So does that mean that robots will eventually be better at all things than humans?"

"Robots," said Botboy, "will make it possible for humans to be better at all things themselves."

The next morning I was released to go home with the superfluous instruction not to drive. The nurse of the day gave me written medication instructions and a prescription. Once she was out of the room Botboy read the instructions and then popped the paper into his mouth. He hated clutter.

We drove home in silence. I was exhausted and woozy from pain medication. Botboy got me comfortable on the couch after he had helped me to the bathroom. Modesty was gone, though it had taken a while to leave. The more like a companion Botboy seemed to me, the more I projected feelings onto him, resulting initially in a sense of embarrassment whenever he handed me a towel as I showered or came to put new toilet paper in the dispenser while I was on the toilet. But I quickly learned that these were my issues. Botby was not human and had no revulsion nor sense of embarrassment programmed into him. It would have been much more uncomfortable to have a human help me in the bathroom, broken arm or no.

I dozed off and awoke to the smell of dinner. I was filled with a sense of comfort and gratitude. I realized that my life would have been very different had I not had this gift from my Rob. What would it be like to be older, injured and alone? Then and there I resolved to make other people aware of the gift to humanity of social robotics. And I fell asleep, sweetly.

11

Once my arm was fully healed, I decided that a first step towards understanding enough of robotics to even begin to act as a spokesperson was to get to know Rob's former graduate students, those who had helped him develop Botboy in the first place. Their silence and apparent lack of interest in their most advanced creation puzzled me a little. They never called to check up on him, nor to offer any information to me.

Botboy suggested we take Faraoun with us to the lab when we visited as he was keen that the boy be able to walk in Rob's shoes, should he ever prove interested and capable. Certainly the robot saw something in the ungainly youth that was invisible to me. But then, I had not known my husband as a boy and had so little contact with his eccentric sister that I was unlikely to ever find out more about what the embryonic Rob had been like.

Said sister had in fact decamped with her husband to the Sahara, on a month-long research project and to my surprise offered Faraoun the chance to go as well. This he refused because of a reluctance to be away from Botboy more than any interest in school, but his parents did not press him and he stayed alone in their empty house as he had done so often before.

Botboy, Faraoun and I drove to Rob's old lab. It was lodged in an unprepossessing building near the university campus, but not directly on it. It had been built in the 1950's and was slowly coming apart at the seams, ivy entering openings where the roof met the walls of wood. The asphalt of the parking lot was cracked and weeds grew up within those cracks. There was only one car parked there and one battered old bicycle leaning against the front of the building. But I knew that the dilapidated wooden building concealed a state-of-the-art surveillance and security system.

We rang the bell. We rang again. Cameras we could not see were doubtless informing the inhabitants who we were because the door was buzzed open after a moment.

I walked down the hall, familiar to me from another life, feelings of sadness and loss and disorientation filled me. It was as if Rob could any moment come around the corner to meet me for lunch, something we used to do every month or so. Instead of Rob, his graduate student, the one named Sirhan, the one who had delivered Botboy to me, came out of a room to greet us. Where was everybody?

He appeared surprised to see us. "Greetings PT 10-06...and Mrs Feenaughty."

"PT 10-06?" Faraoun asked, a little rudely.

"Protoype 10-06," Botboy said, "My designation before you named me."

"Hello Sirhan," I said, thinking he looked thinner than usual and more solemn. "How are you?"

Sirhan gestured around the empty halls, "As you see. I am the only one left. Since Dr Feenaughty died and PT 10-06 was completed and delivered, everyone has moved on."

"Moved on where?" I asked in astonishment. There had once been a dozen young researchers with Rob, physicists, computer designers, robotologists, AI psychologists. Money had flowed into their lab, when Rob was in charge.

"Oh, mostly MIT and Yale... or the government. Once the money ran out and the DOD was headed by anti-robot guys, our funding dried up. I myself am reduced to working on robotic dishwashers."

"A terrible waste," Botboy said.

"You might say that. Dr Feenaughty taught me things of the next century and I am stuck working on things of the last... to make ends meet."

"But," Botboy added, "Those are only some of the things you do. On your own you work on... your old things, do you not?"

"PT 10-06, you always were 'perceptive.' You're right. I still do my own things."

"No longer PT 10-06. I am Botboy." Botboy corrected him firmly.

"OK. Botboy." Sirhan said in a conciliatory tone and grinned as if to say, you create a social robot and then you have to indulge his preferences.

"I came to talk to you about the future of social robotics," I said, though to my ears this sounded ridiculously grandiose. I rushed on with it "My husband would have wanted the world to know about the advantages of a social robot for ordinary people, especially the elderly and the ill. I would like to learn to be a spokesperson for this. But I would like your help, because you are a creator."

Sirhan's eyes showed some light, as if we were now speaking of a dream he had once had but been forced to give up.

"I would like that very much, Mrs Feenaughty."

"Call me Grace please."

"Grace... But there will be resistance," he said, "lots of it."

"I know."

"Chukkerpuppy has already contacted us," Botboy said.

"Chukkerpuppy..." Sirhan said bitterly, "he was responsible for seeing to it that our funding dried out. Religious fanatics within the Administration campaigned relentlessly against public money going to robotics research and egged him on. They believe that working on social robotics is 'playing God.' He knows that is absurd, but as he fears robots for other reasons he plays up to their concern." Sirhan had a melodic and rushed way of speaking, hinting of the Middle East, though his English was perfect.

How did we wind up with a government with so many religious fanatics in charge? I wondered, glad Rob had not lived to see it unfold, to devastate his lab and life's work.

"How is it playing God?" Faraoun asked, edging closer to Sirhan.

Sirhan eyed him, this adolescent scion of the man he had so admired. He seemed to know who he was, "Once a robot can talk and respond to humans on a near-human level two things happen: One, the robot behaves as if it had 'feelings' and its 'own thoughts' and two, human minds are then internally compelled to attribute personality and 'being-ness' to it. This is a projection of course, but an inevitable, unavoidable and very strong one."

He began to wave his hands around, warming to his subject, "And then, then how do you treat such an advanced robot? Like a person? A being with civil rights? If it can have sentience ...feelings... or more likely, can appear to... how does it differ from a man? And is man allowed to create a being, essentially or apparently, a life form, when only God is said to have done it before?"

"Robots do not exactly have feelings," Botboy said slowly, "although a highly verbally competent one may appear to do so, by reflecting back the human's own words and thoughts."

"I know that and you know that," Sirhan said firmly, "but people endow their robotic vacuum cleaners with feelings and project personality onto their cars. They do this automatically, it is wired into the brain of humans, presumably because social networks are humankind's most powerful evolutionary advantage. With the exception of some genetically defective or traumatized ones, all humans are social and create *relationships*. These relationships are far more powerful than they can hope to control consciously." His dark eyes glowed, he leaned towards us on his tiptoes as he spoke. I was glad his opponents could not see him now, the very image of a mad scientist. I almost giggled.

"And speaking of consciously," he went on, "does this mean that a robot can have a soul? What is consciousness? For hundreds of years philosophers and neuro-scientists have struggled with this... this... human attribute... coming ever closer to what they think is an answer... and then along comes a Botboy to challenge it all."

"Can't you think independently, Botboy?" Asked Faraoun, "Do you really just reflect back what I say?"

Botboy hummed a little, as if to emphasize his machine nature, "That is complicated, Faraoun. I have a larger database than you do from which to draw conclusions, yet you can leap to conclusions for unconscious reasons, whereas I cannot. You have intuition that is separate from my abilities. I am, if anything, an extension of your and mostly of She's own abilities."

Botboy put a reassuring gloved hand on the youngtser's arm. I watched as Faraoun relaxed. Rob had often told me that his social robots had been programmed to imitate human gestures such as this, or such as smiling and nodding when appropriate in conversation. He said that when a robot made a small—calculated—gesture such as placing a hand on someone's arm, the human's mind filled in the rest. Beings capable of human-like gestures were naturally and inevitably felt to have feelings. I thought about it. What if Botboy called Faraoun "my child" or "son" when he spoke to him? Faraoun's brain would almost certainly process this as affection, as connectedness. How easily humans could be manipulated by mere imitation of their own ways of speaking.

"That's right," Sirhan said in response to Botboy's explanation, "And what Chukkerpuppy fears, among other things, is that robots are already able to process information more rapidly—and have access to databases far larger—than any human. This obviously has the potential to extend the powers of the human who controls the robot. But Chukkerpuppy worries that robots will control us instead. Dr Feenaughty was alive to these concerns, though he did not share them. That is why, for him, and robotologists like him, the real First Law of Robotics is that Robots may not replicate, as convenient as that would be in the lab."

"Surely you don't mean all robots are made by hand, by people?" I asked.

"No. Simple robots, like self-propelled vacuum cleaners who 'know' where to clean, are made by machines. But Botboy, who would have the ability to design and build another robot, is forbidden to do so. An innate 'stop' has been programmed into him. For better or for worse." And then suddenly, as if remembering his manners: "Would you like some tea? I was just making some when you arrived."

"Yes please," I said, grateful for the implied opportunity to sit down. We followed Sirhan into a nearby room. There were typical lab benches around the edges of the room made of dark colored imitation slate and stacks of electronic parts and tubs of what looked like old-fashioned nuts and bolts and computer parts. It was strange to think of this as the birthplace of my Botboy. Yes, my Botboy. So clearly mine.

Botboy sped ahead of me and opened a folding chair so I could sit at the battered wooden conference table in the middle of the room while Sirhan hunted for cups. Faraoun merely hoisted himself onto one of the lab benches, moved aside the electronic mess and folded his long legs up. He wrapped his arms around his legs comfortably, placed his chin on his knees and looked for all the world like an adolescent preying mantis perched among the detritus of lab science. Nether Botboy nor Sirhan told him to get down, accepting the odd youth without comment. Perhaps odd was a relative concept to them.

Sirhan placed the cups of tea and a bowl of sugar on the table before speaking.

"What are robots?" He asked rhetorically. "Robots are merely extensions of human capabilities, even when they exceed them. Robots can do for us what we wish to do for ourselves. They can be steered by humans, but given how little time there is in a day, they must be autonomous to some degree in order to be truly useful to us. They must stretch our capabilities, be an adventure into the future... They must boldly go where no man has gone before." He smiled at his own joke, "Robots can be any size and of any degree of complexity. They can be simple vacuum cleaners which can sense the level of dirt on the floor and steer around objects. They can be robotic drivers for cars and airplanes, safer than any human as they cannot be distracted, fall asleep, be affected by alcohol or misjudge a distance. They can be like Botboy and have near-human social abilities... And they can be the ultimate life-extenders."

"How is that?" I asked, aware that my joints ached as I sat on the hard chair. How old I felt!

"They can be very, very tiny 'bots... nanobots... which swarm in the bloodstream and do immune surveillance, knocking out cancer cells before they can multiply, reversing autoimmune processes that lead to things like multiple sclerosis or... arthritis. They can, or they could, to a large extent reverse most of the effects of aging. They could repair the heart and blood vessels. The human body is very frail."

No kidding! I remembered Botboy trying to explain this to my horticulture club and Rob trying to explain it to me. It seemed very far-fetched, if desirable.

"As you are aware, this is what is known as body version 2.0, where life is extended and disease and illness virtually unknown. But it requires a leap of faith, a trust in robotics that simply does not exist yet."

I thought about nanobots which I could not see "swarming" inside me and felt a little nauseated. But to be free of arthritis? To live for a very long time? I looked around Rob's lab. Would I like that? What was it worth to live hundreds of years, free of pain, with a semblance of my own former youthful body, but with Rob gone forever? If the man I had adored could not have this life extension, would I really want it? I could not answer that question, the concept was too new and the presence of Rob too strong in the room. I felt tears burn behind my eyes as Sirhan spoke. Could I ever love anyone or anything as much as I had loved Rob, if I lived a hundred years? Did I even want to?

"Sirhan," I said, clearing my throat, "where do you live? I would like the four of us to meet as often as possible. I am serious about promoting Rob's work and I don't care who opposes it."

Where was I getting this sudden conviction? It seemed to be in the air in that old lab smelling of metal parts and dust. Sirhan looked embarrassed, "Uh... these days I guess I pretty much live here. It's... um... hard to make a living in robotics without Dr Feenaughty around." He cast his eyes on the ground, this brilliant young man whose very life had been saved first by missionaries and then by the opportunities opened to him by my Rob.

Faraoun unfolded himself suddenly and shouted "With me! You can live with me! I'm mostly alone in my parents' house. They wouldn't care! They'd like it, actually... so they wouldn't have to make a pretence of coming home to check on me... you can live with me Sirhan! Right Botboy?"

I expected my reasonable robot to tell him he would have to ask his parents, but Botboy only nodded and said, "Yes, certainly. certainly."

And that is how Sirhan came to live with my nephew.

12

Rural Break-in Puzzles Law Enforcement read the headlines of the local paper the next day. I will tell you how it happened.

I had begun the process of corresponding with groups to whom I wanted to make my presentation on social robots. At first I considered the teams at MIT and Yale, at Carnegie Mellon and other well-know AI and Robotics centers. But why preach to the choir? It was not they who needed to be convinced of the usefulness of a social robot to ordinary folks, it was ordinary folks. So I thought I would start—somewhat ironically—with church groups. I had emailed several, of various denominations, beginning with Christians, as they were the most familiar to me. Later, I told myself I would go to Synagogues and Mosques, about which I knew little. And of course to the Baha'is for whom I had a soft spot, remembering their rescue of the child Sirhan.

So I went to bed feeling that I stood on the edge of a new project, one that connected me to Rob but was of my own daring. I felt strong and purposeful. Botboy had made dinner for all three of us: Sirhan, Faraoun and for me. Of course he did not eat with us, for he did not eat, but after the food was on the table he always stood by or "sat" in a chair and joined the conversation.

Naturally we valued his opinion and he seemed to value ours. It was rather like having an extremely bright and very, very logical, rational person at a table of children who are full of ideas, imagination and enthusiasm. He brought that out in us, perhaps because it was not his strong point and so emphasized that it was ours. I have to tell you, I was as enthusiastic as the fifteen-year-old was, a feeling I had not had in many years.

The night had a hint of Fall to it, cooler by many degrees than the hot day, with a smell of leaves and fungi from the forest. I breathed deeply and feel asleep almost at once.

I'm not sure what awakened me, I imagine it was a noise. Occasionally in the early evening, after I had gone to bed, Botboy would putter about the house picking things up, arranging dishes for the next day, attending to the minor tasks still remaining to him. The sound of him moving about was comforting. But as I lay there in bed I knew this was not early evening. It was pitch dark and there was no bird song and no sense of early dawn. My guess was that it was one or two in the morning and I lay there oddly uneasy, straining to hear again what had awakened me in the first place.

I could not explain why my heart was beating fast and why my arm hairs stood on end. Rob always claimed that the brain, even in the barely conscious interface between sleep and awakening, processed information and drew conclusions it would take the slower rational part of the mind some time to catch up with. What had I heard and where was Botboy?

There it was again. Zzzzt. It sounded like the sound you would make if you took a sharp knife and slit a metal screen. I hoped I was wrong. I sat up. Zzzzt. Zzzzt.

Suddenly Botboy was beside me, leaning his large metal head close to my own. There was no humming or beeping.

"She." He whispered very, very softly, something I had never heard him do. "Get in the closet and stay there."

I didn't need to be told twice. The snick of the door closing behind me as I crouched among the hanging dresses and robes sounded as loud as a firecracker in my ears. My heart was pumping blood madly and my mouth was dry.

The closet door was louvered and there was some ambient light from the sliver of moon, once my eyes adjusted. I could see down the short hall from my bedroom to my bathroom. The sound seemed to have come from there. Was someone entering the house through the large window to my bathroom? I usually left the window cranked open to let in the sweet night air.

The bathroom was on the side of the house that looked out across the fields and away from the automatically lit-up front entrance. As far as the eye could see, there were no neighboring farmhouses in that direction. About 100 feet away was an old hedgerow offering cover. It was thick with maple and fir and Amelanchier bushes, a warren offering rabbits, skunks, coyotes and squirrels their home. I had often imagined, before Botboy came, that if threatened by an intruder, I would seek refuge in the hedgerow myself. Of course I never stopped to think of the absurdity of an older lady in her pajamas, small revolver in hand, sprinting across the plowed field towards the hedgerow with its uncertain footing. In my mind, when threatened, I had wings on my feet, like in dreams, and could run as swiftly as a girl. And who knows? When sufficiently threatened perhaps even my arthritic bones, feeling the flash of adrenaline, could run and jump and bend to hide. Survival.

My gun. It was in the bathroom underneath the folded-down ironing board. I certainly had no way to get to it. Whatever intruder was coming in through my bathroom was closer to it than I.

Botboy did not enter the bathroom. Rather he stood as still, as immovable, as a pillar just outside the door. Whoever came in through the bathroom would have to pass him. Likely, in the dark they would mistake him for a piece of furniture. I remembered his words to me: *No human is as strong as a robot.* And prayed he was right.

But humans with weapons?

It was clear now to my ears what was happening. The sound, however soft, of feet hitting the floor was unmistakable. Then another. There were two of them. Botboy did not move at all in the thin moonlight. I was terrified.

There was silence for what seemed forever. I knew that all my senses were on fire and that with the slightest provocation I would begin to scream. Only the thought that Botboy remained silent and clearly expected me to do so as well, kept me sanely immobile, my teeth clenched trying to control my breathing.

Then the soft, soft sound of footfall. They emerged from the wide sliding bathroom door which I customarily left open when I was not in it. I could see they wore ski masks. One had a backpack and the other carried what looked like a weapon, a long wicked weapon, like pictures of an AK 47 in old Vietnam War movies.

What happened next was so fast and so brutal I did not realize I was screaming until it was over. Botboy reared up in the dark, looking twice his size, green light shot like beacons from his eyes and he too began to scream, only in a high-pitched almost female voice, like a samurai. It was the most terrifying thing I had ever heard or seen and the intruders thought so, too. Faster than any of us could comprehend it, Botboy twisted the gun from the lead intruder's hand and threw it in my direction. It skittered across the hard concrete floor and landed under the bed, deep under the bed. The intruders made no attempt to go after it. Botboy came at them as they leapt backwards and scrambled over the low frame of the large, half open window in my bathroom. I could hear the sound of shattering glass as my antique porcelain Chinese vases and Bodhisattvas were knocked over and destroyed. The intruders did not utter a word, did not scream or yell and made no attempt—wisely—to stand and challenge Botboy. He let them get away. I know he did. The same hands that could file and twist and lift heavy weights could easily have crushed a skull, but he did not do it.

I was out of the closet in a flash and at his side, foolishly perhaps, but instinctively. We watched the two terrified men run up the hill towards the road, literally as fast as they could go. In the distance I heard a siren. Botboy must have called 911. I became aware of his speaking voice, his normal speaking voice, giving directions on the cell phone that he was himself.

"They are running towards the road," he told someone, hopefully the Sheriff. "The road runs North/South and there are few homes in either direction." Then, "We will wait for you."

I felt weak at the knees. I put a hand out and Botboy steadied me, but he did not take his eyes from the retreating figures.

"What if they go get guns and come back?" I asked him, my worst nightmare of the many I had on those scared nights after Rob's death.

"They will not be back," he said, "at least not for a while...She," he turned to me, "The Sheriff will be here in about 7 minutes. Perhaps you want to get dressed?"

The Sheriff's tires crunched against the gravel driveway, doors slammed. Before they could ring the bell Botboy opened the door to them, I could hear them speaking on their mobile phones about back-up. The flashing light from their cruiser lit up the countryside surreally. The crackle of police phones cut through the night.

"Cripes!" Yelled the Sheriff, catching sight of Botboy and involuntarily backing up a pace.

"It's just my robot." I hurried to say, "please come in."

It took a few minutes for them to be comfortable with my explanations about Botboy, who he was and what he was capable of. Still the younger deputy eyed him mistrustfully.

We showed them where the intruders had entered. The Sheriff closed the door after briefly inspecting the cut screens and the shards of my antiques.

"Ma'am, I need you to not enter this room again until we have examined it." He said firmly. "We'll dust for fingerprints."

"There aren't any," Botboy said matter-of-factly.

The Sheriff looked at him intently for a minute, "We'll see about that. Where's the weapon?"

"Under the bed," I said, "where my robot tossed it when he... disarmed them."

The Sheriff nodded, as if considering this, and then jerked a thumb at his deputy. The younger man knelt at the bed, but finally had to lie full length to reach the weapon. He withdrew it carefully. Close up it looked slightly different than I had thought, all hard black plastic. Was it real?

"AK 74 M," Botboy said, "side-folding plastic buttstock and the scope mounting rail on the left side of the receiver. Russian Army issue. Older gun."

The Sheriff just grunted and removed a large old-fashioned white handkerchief from his pocket. The deputy handed him the weapon and the older man took it gingerly, protecting it from fingerprint contamination with the cloth.

"Freaky weapon for a home break-in," the deputy said in awe.

"You might say," the Sheriff said, "even for druggies, odd. Very odd."

"I filmed them." Botboy said.

"You what?" Despite his obvious attempt to appear cool and unsurprised around the robot, the Sheriff stared.

Botboy handed him a disc which he had ejected from somewhere on his metallic body. I had seen him do this before, most recently when he filmed the little fox marauder for me. But the officers were speechless.

"Handy robot," the younger man finally said with emphasis, "Wish I had one. Wish the department did."

"Soon, perhaps very soon, most people will have their own," Botboy said kindly.

"Mrs Feenaughty," the Sheriff said, "What do you have of value in this house that someone would want?"

I avoided saying that Botboy was my most valuable possession, by far. I shook my head, "I don't even have a television to speak of... No cash... well some... but only perhaps a thousand dollars in my laundry basket." Boy did I feel like a dumb old lady from the previous century saying that!

I went on, "We have a very large book collection, some of them rare. Some art. Not much. Mostly my husband's own."

He shook his head. "Your robot seems to have given them the fright of their life, but I wouldn't count on them being ordinary intruders. The lab will be out in a few hours to go over the crime scene and interview you further. *Why* will be important to know, if we can." He looked pointedly at the robot. Botboy smiled pleasantly but said nothing.

After a little while I showed them to the door. Botboy insisted I lie down and sleep, but I did not go near my bedroom again. Stretched out on the long leather couch I was surprised the next morning, when I awoke late from an amazingly dreamless sleep. I sat bolt upright. I could see Botboy in the kitchen drying glasses with a cloth, reluctant to use the dishwasher because of it carbon footprint.

"OK, Botboy," I said, "What's really going on?"

13

Before Botboy could speak the front door opened and both Faraoun and Sirhan came charging in, speaking over the top of each other, firing questions at both me and the robot, whoever might answer them first.

"What happened? Are you alright? Why didn't you call us? Who were they? What did they take? What were they after?"

"That's what I would like to know," I said, "And Botboy, there's something you're not telling me, I just know it."

Botboy placed coffee mugs on the table and began to pour precise aliquots of coffee into each, for Sirhan he had made tea.

"I do not think the intruders came to rob She," he said slowly.

"Did they come to kill her?" Faraoun asked tactlessly. I glared at him.

"I do not think so. Their weapon was formidable enough, but I think it was intended for me or whatever security Founder may have placed in wait for them."

Faraoun leaned over, spilling his coffee, "Wow! What if they had shot you, Botboy, would you have died?"

Botboy righted the boy's cup without reproach and wiped his spill with a napkin. But he did not give him more.

"I assume you mean would it have destroyed me, Faraoun. I cannot exactly die. But the answer is no. It might have created a very dangerous situation to nearby humans, though, including the intruders themselves, because of ricochet. Therefore it was of first order importance that I disarm them."

"But you let them escape, Botboy," I protested. "You could have killed them or injured them."

"He had to, Aunt Grace," insisted Faraoun loudly, excitedly, "It's the First law of Robotics! I've been reading about it. Isaac Asimov: *A robot may not injure a human being or, through inaction, allow a human being to come to harm.* Isn't that right Botboy? Isn't that right?"

Sirhan put a hand on the young boy's sleeve, like an older indulgent brother,

"That was fiction, Faraoun. It's a nice idea, but highly impractical, kind of like saying 'all children will obey their parents.' It had to be considered and then discarded as much too simplistic even for the early days of advanced robotics."

"But then... then robots are much stronger than we are. They could just kill us, right?"

"Social robots like Botboy would never kill you, Faraoun, because you are their family, their tribe, their... their pack. You have been adopted, so to speak, just like a dog will adopt you as a member of its pack. Your own dog will not kill you. And then, of course, a moral sense—and by that I mean a sense of right and wrong—has been programmed into your robot."

"Right and wrong," Botboy repeated firmly.

Sirhan looked at him steadily, "That does not mean that we will not need to come up with a set of laws that govern human-robot and robot-robot interaction. Imagine human life without laws that govern human-human interaction. And we will need to be thoughtful and deliberate, for as you point out robots are much stronger than we are and—eventually—they will be immeasurably smarter. In a way, they will go from being our children to being our parents. But if done right, the same bond of what we humans call love or connection will exist either way."

Faraoun sat back in his chair with a thump.

Why must that child be so physical?

"So... you could have killed them, Botboy. But you didn't. Why not?"

"Killing is wrong and to be avoided in almost all conceivable circumstances." Botboy answered. "'Thou shalt not kill' Exodus 20:13. An early piece of your own powerful social conditioning."

"Honored more in the breach..." Sirhan muttered, thinking of his own childhood no doubt.

"So you didn't need to kill them to protect me..." I said, "but couldn't you have captured them?"

"I miscalculated their fear. Did you not see how frightened they were? Had I struggled with one of them, I am so strong his arm might have come off in my hands. Then we would have had a real emergency. And a mess."

I felt lightheaded at the prospect.

"No, She, in retrospect I should have remained very still and waited until they walked right into me, unsuspectingly, and then simply caught the one with the gun. The other would have run away, but we would not be left guessing. We would know why they came."

"You made a mistake?" Faraoun asked wonderingly, "Robots can make mistakes?"

"Of course," Botboy answered, "I will learn from this mistake and I will not make it twice. Some robots are smart precisely because they can learn from their mistakes. Like humans."

"Only without the ego," Sirhan said.

"But this is not about Botboy," I protested, "It's about why two men armed with a military assault weapon, 'Russian issue' as Botboy says, came in through my window at night!"

"Oh, but I think it is about Botboy," Sirhan said. "He is surely the single most valuable thing in this house... or most other houses." He turned from me to the robot, "Could they have wanted to remove you?"

Botboy thought for a minute, "I doubt it Sirhan," he said. It was a peculiarity of Botboy's that he pronounced the young man's name as if it were a title: "Sir Han." Perhaps it was an inside joke? "If that had been their intention, they came poorly prepared indeed. No... I think they wanted something else. I imagine the weapon was to threaten She with, to insure that I would not interfere. A high stakes gamble indeed. You notice, She, that they did not have a car parked nearby, so removing something large would have been impossible. And one of them wore a light backpack. So they wanted something small. They wore gloves so there would be no fingerprints, but despite their acute fear they did not speak or even scream, so they left no voiceprint for me to analyze. They are, therefore, sophisticated. I got their picture but they wore masks."

"Something small?" I asked, "Whatever do I have that is small and valuable?"

The robot turned to Sirhan, "Sir Han, what did you do with Founder's notebooks?"

Sirhan thought for a minute, "The short answer is: Nothing. Dr Feenaughty left only a few notes around the lab. His real notes—taken by hand, not on his word processor—were in some small notebook he always kept on his person. I assumed it was with him when he died."

Oh God. Rob had been cremated at his own wish. Had I allowed his notebook to go up in flames with his earthly remains? For a moment I couldn't breathe. They were all looking at me. Then: Of course not. Of course I hadn't. Rob had been in his pajamas when his heart attack overtook him and I knew for a fact he did not sleep with a notebook on his person or anywhere near him. And I had buried him in his wedding suit, the dark green one that, incredibly, still fit, and which he hadn't worn since that day in San Francisco where we had exchanged the vows that bound us for the scope of our whole lives... his life. Ah God! The pain of it.

"I have never been through the library," I said when I could trust myself to speak.

"Neither have I," said Botboy.

"Let's do it!" Faraoun shouted, excited by the adventure. I could have killed him.

"Faraoun," said Sirhan reproachfully, "This is not easy for your Aunt. Please do not shout. Mrs Feenaughty, it does seem as if we should go through the library very carefully. Those notebooks are priceless scientific work. One day they will be priceless historic work. May we?"

"We will do it together," I said.

And we began after breakfast. Faraoun was so excited he could hardly eat.

"He is a child," Botboy said to me quietly.

We decided to go over every inch of the library together, as inefficient as that sounds. Four sets of eyes can see what one may overlook, especially given the differences in viewpoint: That of a child (sometimes the most powerful viewpoint of all), that of a scientist, that of the person who had loved the creator of the objects we sought, and that of his Robot.

Rob and I had collected books all of our lives, separately and then together once we were married. He collected scientific articles and early books and I collected Charles Darwin and books on butterflies. Oh yes, and novels by Nabokov, a great lepidopterist, and novels by Nayantara Saghal, obscure Indian author and niece of Nehru.

There were at least 10,000 volumes in our large room and they lined the walls to the ceiling. We had two desks, his and mine, as well as stacks of books beside our armchairs, his untouched since that day. There was an old microscope of my father's and a school dissecting scope for my work with butterflies. The many volumes, no longer dusty thanks to Botboy, looked down on us like benign friends, each with its own story to tell: The story of its contents of course, but as well as the story of its acquisition and what acquiring it had meant to us. So there, in books, was the story of our life together, Grace and Rob, encoded in the accumulation of all the volumes.

Where to begin? There were no carpets to have to lift, like in old detective stories. And a glance through the blueprints from the original building plans revealed no hiding places. There was no safe. That I knew. We did look in every Chinese pot large enough to hold more than a matchbox. Nothing. Then we searched his desk, one person—in this case me—doing the physical searching and the others looking on. Rob kept many things in his desk as well as pictures of us, of me, of Faraoun as a baby and an odd assortment of photos of his co-workers from the lab. A sad silence descended on us as we looked at his things.

"Uncle Rob," Faraoun said softly, touching his Uncle's furrowed face in a close-up photo.

"Dr Feenaughty," Sirhan said sadly.

Rob, my heart said, Rob, Rob.

"Founder," Botboy said, finally.

He turned over a sheaf of papers that looked to be the start of an article. Rob put an amazing amount of information on paper, considering we lived in the age of computers. But we found nothing in the desk and moved on to the books, a little daunted. After an hour or so, Botboy stood to one side to think. He did not speak for some time as the two younger men and I went methodically through each book and peered behind it. Botboy was trying to puzzle out where the most likely, most logical, most Rob place to put something could be.

Suddenly he looked up, "She. Of course."

"What Botboy?" I asked.

"Search your own desk." He said, "I am a gift to you. Everything Founder left he left for you."

I almost dropped the book I was holding on the Flora and Fauna of Kenya. I raced to the desk and began opening drawers. But I did not have far to look. There they were. His small black notebooks, held together with a rubber band, filled with his close, somewhat crabbed writing.

I picked them up. Sirhan reached out a reverent hand to touch them, Faraoun's eyes glowed.

"With your permission we should go over them carefully," Sirhan said, "It will take weeks, of course."

I clutched them to me, "Yes. Yes. We will go over them. But in the meantime where will we keep them safe?"

Botboy chuckled. Can you imagine a robot chuckling? It was very strange. He looked at Sirhan, who smiled.

"If you will trust me, your robot, She, I will keep them safe. Touch my chest below the small label, if you would."

I looked. There was a small metal label with some numbers on it that I had never really noticed before. I touched the area below it. A small drawer slid out.

"It was made to open at Founder's touch or yours. Now your touch alone can open it, She. Neither Sir Han nor even I, much less another, would be able to access it. Can you bear to place Founder's books in here?"

I had to laugh, despite the pathos. "First you are a cell phone, sometimes you're Jeeves, you are several computers at once, at least one camera and a carbon recycler. And now you are a safe. What next?"

I handed him the small books and he placed them in the drawer. He pushed against it and it closed soundlessly. No seam was visible.

What next?

14

Botboy augmented our alarm system with a few improvements of his own. He researched wind turbines and solar panels, deciding finally on the latter, despite the expense, in order to avoid a long stretch of wiring from the turbine to the converters and from there to the house. He wanted the alarm system to be independent of electricity from the road, where it was vulnerable to intentional disruption. He calculated that with a few energy saving efforts such as turning off all appliances when not in direct use and hanging the laundry to dry, we could be energy independent. But then he realized that he had forgotten the plug-in hybrid car he had talked me into ordering for delivery later that year and had to expand his choice of panels. Perhaps, he suggested, we might need to opt for both solar and wind. Time would tell.

"We must get off of hydrocarbon usage, She," he said firmly "Founder left you with enough money to afford infrastructure improvements and we should take advantage of that. After all, I calculate your lifespan to be..."

I put my hands over my ears "Stop! Don't tell me," I said. I had no wish to know how many years without my Rob the insurance mathematicians thought me good for. It was one thing to cherish each day as it came, despite my loss, but I could not contemplate long stretches of my life without his companionship. It was irrational, but real.

Rob will never return, a voice inside my head told me during black moments of anguish. I knew it, of course, I was not a child, but I couldn't—yet—see myself accepting it.

Botboy and I both derived a little morbid pleasure from thinking about the fact that whoever had tried to steal Rob's notebooks was probably still trying to figure out how to get into the house, meanwhile my robot had the object of their efforts on his person—in his person to be more accurate—the whole time we were at the grocery store, in the garden or driving to give talks at a church.

We got our initiation in spreading the word about social robots by talking to local church groups. But it soon became clear that our part of the country, immersed as it is in the high tech industry and in the production of alternative energy and all the technologies that go along with it, was far from being representative of the country at large. In our city one met many people who had robotic cars, robotic vacuum cleaners and car washers. There were even robotic security guards, who patrolled the perimeters of buildings and reported suspicious activities via an internet connection or cell phone. Children played chess and checkers against robots. And although no one had a social robot as advanced, or anything like as advanced, as Botboy, people in general were not uncomfortable with the concept. The advantages were too obvious. No, the challenge Botboy, Sirhan and I must face would be in towns and cities far from the Techo-hubs of the West Coast or Northeast seaboard.

Faraoun pleaded to go with us to spread the word on social robotics. But this would have required him to miss too much school. Sirhan had simpler robotic work to continue doing at the Lab and so often stayed to be around the boy. My sister-in-law and her husband, as predicted, had been delighted with a low-maintenance live-in caretaker for the child whom they ignored anyway. They celebrated by spending most of the year in the Sinai.

This meant that I had to go alone, and by plane, to visit churches in Omaha, Ogden, Tulsa and Missoula, places I knew only from geography lessons and—more rarely—from the daily news.

Well, not quite alone. I had Botboy.

Botboy packed my small suitcase for me, having first carefully researched the airlines regulations about carry-ons. Security had one set of rules, which we followed as a matter of course, but the weight regulations were the result of efforts to save jet fuel and so Botboy insured that we never, ever exceeded the allowable amount, even though I could have paid the excess baggage charge. Far be it from my eco-frugal robot to allow consumption of any more fuel than absolutely necessary.

Botboy spent hours on the phone, pretending to be my secretary, in order to find out whether or not he could buy his own plane seat, rather than be forced into the baggage hold, something we both would have found completely unacceptable. Over the phone people responded to him like another person, thinking he was one. Robots like him had long ago passed the Turing test: Their conversation was not distinguishable from that of a person's (a smart person's). The airlines finally allowed me to buy an extra seat, never thinking, I suppose, that Botboy could be one giant cell phone that could not be switched off.

Security that first trip to Missoula, Montana was a nightmare. We missed our flight. We rolled quietly through the long lines without incident and only a moderate amount of pointing and staring. Botboy had made an identity card for himself, stating: ROBOT MACHINE: Property of Grace Feenaughty, my address, phone number and passport number. The word "machine" had been inserted to make him seem less scary. He had placed his ticket, face up, in a plastic holder hanging from a lanyard so as to be easy for the security personnel to read. He carried nothing. We had agreed that he would not speak unless necessary and then only in his most machine-like voice. I would give him commands in a clear, slow voice. I hoped I would not laugh.

The first security guard did a double take as we approached him.

"This is my robot," I said before he could speak, "He has his own seat."

Botboy held out his card for the man to read.

"Get a supervisor!" The man called out loudly, "Now!"

People moved away from us as if from a bomb threat. Did they imagine that a real threat would merely initiate a call for a supervisor?

"It's a wobot," a little girl said in awe, but her mother grabbed her thin arms and frog-marched her away as fast as she could go.

A somber, grey-haired man in an official TSA uniform and hat approached us. He took my I.D. and ticket,

"Mrs Feenaughty," he said, "You can't just take a robot on board an airplane without inspection."

"The airlines knew he was coming," I said nervously. What did they mean "inspection?"

"Step over here," he said, and then to the crowd firmly: "Nothing to see folks. Nothing to be alarmed about. Just go through security." But no one moved. They were amazed at the sight of Botboy standing quietly one minute and rolling off in the direction the guard indicated the next.

"Call for the canine team," the guard said to a young, open-mouthed, dred-locked colleague at his side, who turned and ran.

"Remove your shoes please Mrs Feenaughty and step over here."

Two women wanded me without event, for Botboy had insisted I remove everything remotely metallic before we left the house. My sole carry-on, a purse, was opened, ransacked with scarcely concealed delight, each piece carefully wanded. I guess they thought if you hung out with a robot you could be a dangerous subversive. The dog handler was beside us in a flash.

Botboy was of no more interest to the dog than a large metallic table or chair. He certainly was not "alive" in the dog's eyes, he had no human or animal smell. I knew that Botboy was as capable as the dog of sniffing out explosives and contraband, but of course we did not mention this. The dog handler gave his dogs a command and they sniffed the whole area, my purse, my feet and hands, but ignored my robot.

"Nothing," The dog handler told the guard after a second.

The guard placed my purse on the conveyor belt and indicated that Botboy should lie down behind it to go through the detector.

"But..." I began, but Botboy clambered up onto the moving belt and lay down without protest. They moved the purse through, then backed it up a bit and began to move him toward the mouth of the opening. But Botboy was too big, even lying on his "back," to fit. He had placed his "feet" first and the bottom half of him slid through without a problem, but he got stuck at his torso, his head still out, a calm expression on his face. I was ready to scream.

"Robot PT 10-06 zero point five centimeters too big around to enter," he intoned in his machine voice.

A gasp went up from the onlookers "He can talk!" "The robot can speak!"

"He has good speech recognition," I told the guard hastily. "Most of the time." *Sorry Botboy!*

The guard, who must have once been a military man judging from his ramrod straight bearing, close-cut grey hair and refusal to appear surprised, said tersely, "Then tell it to get down and go through the gate over there."

The conveyer belt moved backwards to release him,

"Get down," I said loudly and clearly, as if to a computer with slightly faulty speech recognition. "Come with me."

Botboy did as he was told without speaking and rolled through the metal detector I indicated by pointing with my finger, a pleasant, sweet smile on his face. I had a childish urge to yell, "run!" But of course did not.

Now naturally, Botboy is made of several kinds of metal and so the detector began to scream. We were converged upon by more guards and conducted hastily to a side room.

I sat in a hard wooden chair, my robot standing quietly at my side. A young black woman brought in my purse and gave it to me. She smiled at Botboy and he smiled back.

"Mrs Feenaughty," the older man said, "What made you think you could bring a robot onto the plane?"

I sighed, "You see, I cleared it with my airlines first. I am an older woman with severe arthritis and Bot—er... my robot is necessary to carry things and drive me around in a strange town. He's... he's like a seeing-eye dog."

"Well, seeing-eye dogs have permits and licenses," the man said.

"And they are also alive," I replied. "They are living beings who may bite people, get sick on the plane or run around. A robot is a machine like your computer," I indicated the computer on his desk. *Sorry Botboy!* "It only does what you tell it to."

"The bomb sniffing dogs passed him, Chief," a younger man said helpfully.

"Where are you traveling to?" The older man asked me, ignoring him.

"To Missoula. I am to give a talk tomorrow afternoon at the First Church of Nazarene Disciples downtown. A talk about social robots—which this one is—and how they can help the elderly and infirm."

"Well, you've missed your flight. The next one doesn't leave for an hour. I will escort you to the gate. But I'm warning you that the pilot will have the final say about whether or not he will let a robot this... this... big onto his plane. And for the future I think you had better get special papers from TSA to let you pass without incident."

Seething, I thanked him. Botboy wisely said nothing.

The pilot did come out of the plane after everyone had boarded; I had been instructed to get on last.

"Mrs Feenaughty," he said kindly, "I'm captain Roberts. I understand you have a handicap robot."

I didn't contradict him.

"I'm curious as to what he can do."

"He is a very advanced robot, Captain," I said. "However he does what I instruct him to and no more. He has good speech recognition and can talk to you, if you like."

The captain smiled. "Hello Robot."

"Hello," Botboy answered in his machine voice.

"Welcome aboard Flight 234 to Missoula."

"Flight 234," Botboy repeated, "Scheduled to land in approximately two hours three minutes, if headwinds permit."

"Right you are. Come with me."

We followed him to the first class seats we had wisely chosen. The area was almost empty, though coach was full.

Botboy was given the seat by the window and the captain indicated he should sit there. Botboy glided in, had some difficulty lowering the seat and altering his middle region to "sit" in it at the same time. The captain bent over him and fastened his seat belt.

"Can't have a large metal object like yourself flying around the cabin in turbulent weather!" He said cheerfully, "My son goes to MIT and he tells me robots are the wave of the future."

I smiled. That was a stroke of good fortune. What if his son had studied ancient Roman history? He might have been less sanguine.

The plane lifted off fifteen minutes later. I ordered a glass of wine and Botboy sat with his large eyes closed. I knew he was both trying to be less scary to the parade of people who suddenly "had" to use the restroom up front and trying to analyze the sounds of the machine's engines.

I fell asleep and woke up in Missoula.

15

Missoula is not a large town. It is located on the Clark Fork River in a mountainous valley and has about 50,000 inhabitants, most of them...well...God fearing. We were not scheduled to speak at the Church until all services were over for the morning, for in that part of the world many families spent the entire day in church. We were part of the afternoon speakers to be followed, apparently, by more praying, hymn singing and giving of thanks. Not a bad way to spend a day, especially with children.

Botboy and I were amusing ourselves by taking a walk along the river's edge.

The Clark Fork River meandered across the Autumn landscape like a snake lying in the sun, of which there was precious little that day. I had wrapped myself in a long down coat and scarf; Botboy was unaffected by this weather as by all weather, and walked along beside me, cheerful and happy to observe countryside about which he had only read. There were willows, cottonwoods and pines along the bank and flat sand islands where the water slowed and divided into slow-moving, lazy fingers of icy brown. He identified birds as we walked.

"'Meadow lark,' 'mallard,' 'barn swallow,'" sparing me their Latin names, which I had no doubt he knew or could at least access.

"'Queen Anne's Lace,' 'Dandelion,' 'Wild grape,'" I recited as we passed these plants. He nodded.

The ground was rough and I was pleased that I did not stumble. Since Botboy had arrived I had spent much more time walking outdoors as I was less afraid. Just as you might imagine, the more I walked on uneven ground, the better my balance had become and the lower my blood pressure. Botboy carried a small picnic basket for me with some warm coffee and French pastries he had baked at home. Oddly, I had also found that my weight had stabilized. I ate better with my Robot as company, more nutritiously and less erratically than before. More vegetables. I had been in serious danger of living on tea and toast, something eldercare doctors are always warning against.

We had left River Road and were skirting the edge of the water itself. I found a large flat rock and sat down, Botboy stood beside me and passed me a warm cup of coffee.

He looked around us: "Swainson's hawk," he said, then calmly, "Rattlesnake."

Rattlesnake!! I leapt up onto the top of the rock, knocking my coffee cup into the dry grass. "Where?" I cried.

Botboy pointed with his gloved hands, "There in that rock pile, She. They like the warmth of the sun, too. If you do not tread on them, they would strike at me first as my metal surfaces would be warmer than your flesh. It is an old prospector's trick to get them to strike at a warmed shovel. They strike at warm objects."

The things he knew! I shook my head. Please imagine a funny-looking heap of metal, standing beside a Montana river and talking about "old prospector's tricks"... sometimes it was hilarious.

"Great," I said nervously, letting him retrieve the cup. Who knew what waited in the tall brown whispery grass. "Let's go back. It's almost two o'clock and we have to start at three."

Moving rather more swiftly than before it only took us a half an hour to reach the church. We walked across the lawn that ran to the water's edge towards a building that could only be described as imposing. Who would have thought that such an enormous church would serve such a small community, and doubtless there were other churches as well. No small clapboard structure, the imposing building was concrete and what looked like redwood, a beautiful reddish hue. There were stained glass windows in the front, fully twenty feet high and sparkling like jewels. A mammoth cross was affixed to the prow of the narthex. Hundreds of cars filled the parking lot. There was a baseball diamond and a basketball court to one side. These people were serious about spending the day here. I had declined lunch; memories of jello salad and tuna noodle casserole encountered at my relatives' church on my return to from a childhood spent in France still lingered. But judging from the smells that emanated from a wood fire somewhere out back, I should have reconsidered.

"America," I whispered to myself, not without wonderment.

As we approached a tall man broke away from the crowd at the door. You will think I am exaggerating when I tell you that he wore a large, white cowboy hat and a string tie adorned with a huge, irregularly-shaped piece of turquoise. But it's true. Also true was that he had the kindest, most intelligent face I had seen in many years and a voice that boomed like Lyndon Johnson's of old.

"Mrs Feenaughty! Grace!" He roared and galloped towards us. Botboy stopped in his tracks and observed him guardedly.

He drew up panting, removed his hat and stuck out a hand. "And this must be your robot."

When he had done with my hand, pumping it up and down vigorously, he shook Botboy's. "So glad you chose our little church to preach to," he said with a twinkle.

I could hardly deny that I would be preaching... though not to the choir, I thought.

"Thank you Reverend Smith," I said, "Shall we follow you?"

We followed him, and as we moved thorough the crowd he called out to them as if to favorite children and the crowd responded by following him in a giant wave through the doors. Most people were very well dressed, though casual, hardly the Catholic crowd of my youth, and they chatted with one another happily and openly, no hushed voices, no air of sanctity.

He led us to the podium; there must have been three hundred people seated in the large beautifully appointed nave. He pinned a small microphone to my lapel and looked questioningly at Botboy, "Should I give him one, too?"

I shook my head, "He'll just turn up his volume."

Reverend Smith did not need to call for order. He raised his hands above his head and all fell silent.

"Folks," he boomed, "Let us pray."

"Lord God, God of Abraham, of Issac and Moses, of all your children gathered here, hear our prayer. Help us understand our sister Grace and help her to understand us. Show us the way. Show us your way. Keep Lucifer from our halls, silence his voice in our discourse. Give us a Christian conversation in our Christian home."

I peered sideways at Botboy, knowing he was filing this away for future reference, completing his own personal database on American Protestant evangelism. I could just see his file: Religion: Twenty-first century: Evangelical: Protestant: Mountain West. He would share it with Sirhan and Faraoun and they would try and picture it. I wondered if the database could capture nuances as subtle and meaningful as the cowboy hat.

"Keep the Peace within our House, Lord. Keep thy peace,"

I no sooner had just reflected that this was a wonderful sentiment when he boomed: "Amen!" And the crowd roared back the same.

"Amen! Amen!" Said Botboy, much too loudly. I could have wrung his tinny little neck. But the crowd roared its approval and Reverend Smith turned bright blue eyes on me, "Now that's my kind of robot!"

"Grace," he then said, his naturally resonant voice greatly magnified through the almost invisible microphone, "Tell us about your robot."

I had prepared a speech about social robots, but in this atmosphere it seemed silly and stilted. I drew a breath. These were real people, people who cared enough to learn about robots, to wonder what they might have to offer and to let me share their precious Sunday hours.

"My name is Grace Feenaughty and I am 63 years old. I am a widow. My husband Rob was a scientist... a robotologist... And he designed a robot to take care of me when he no longer could. And that is what you see before you: An older lady and her social robot." I paused. No one took their eyes off of us. "His name is Botboy and he is my true companion, he cooks for me, he drives me to the doctor, he cleans the house, he helps me garden, he called 911 when I fell off a ladder..." I looked around again to see how they took the word *companion*. Mouths were open in astonishment, people were leaning forward in their seats, "Botboy can speak, he can think independently, although he always does what I tell him to, unless it endangers me."

"Botboy," Reverend Smith said gently, as if to a child, "Speak to me, would you, son. How old are you?"

"I was constructed five years ago after a professional robotologist's lifetime of research." Botboy said in a clear, not-quite-human voice.

The crowd gasped. It was one thing to say he could speak, it was another thing to hear how naturally it came out.

"Botboy, are you a Christian?" Someone called out.

"I am not a human," Botboy said, "I am a machine. Therefore, I think you will agree that I cannot be a Christian."

Oh great. Now he was going to start a theological debate. Good luck debating with a being who has instant access to anything ever written on the subject. How about sticking to housework?

"Can you scrub floors?" Called a woman.

"Of course."

"Wash windows?"

"Of course. If they are very tall I may need to employ a ladder."

"Can you baby sit?"

Botboy smiled his most engaging smile at this question. It is amazing what a smile will do for humans. *Smiles and eye contact*, Rob always said, *gets them every time*. "I love to baby sit," he said. "I help Faraoun with his homework, he is fifteen. He hates homework."

Laughter broke out, then, "I want one of those!"

The women seemed mostly interested in what household tasks Botboy could do and the men in what he could repair. Botboy showed them his multipurpose hands, always good for amazing a crowd. But the children asked the most penetrating questions, the adolescent boys largely concerned with the exciting possibilities of robots run amok and of their usefulness in war and wargames.

"War is bad." Botboy said flatly. "An advanced robot like myself should use all of his powers to help humans understand how to avoid wars."

"But can you love?" Asked a little girl from the front row.

Botboy leaned forward. "You are a human. I do not need to tell you what love is, you know what it is...instinctively, innately. I was made specifically for She." He indicated me. "Her welfare is my chief concern. If something is good for her or merely makes her happy, then I am all for that thing. Is that love? I think it might be."

Reverend Smith seemed sunken in thought as my robot talked on, then suddenly he roused himself and asked,

"So you say you can talk and think independently. You can respond so well that another human cannot detect that you are not, in fact, human. You engage in caring for people like human beings do, call it love, if you like. You can plan ahead. You can discern and respond to emotions...then do you have a soul?"

"I know many things," Botboy said simply, "But I do not know what a soul is. Can you define it for me?"

There was silence.

"A soul is that breath of life beyond mere life on earth, that piece of the everlasting which God breathed into man as he created him...in his own image," the Reverend said.

Now Rob had told me many times that man had created God in his own image and hence the resemblance, not the other way around. Botboy knew this, of course, as well as I did, and his deferral to his Founder's wisdom was absolute. But never did my robot seem more human than at that moment, when he held his own counsel and did not speak offensive words out of kindness and respect for the belief of others.

"If, as the Bible says," Botboy said after a moment, "God created the heavens and the earth, the light, the firmament, the fish and fowls, the cattle, wild beasts and creeping things... if he created the heavens and stretched them out, spread out the earth and its offspring, gave breath to the people on it and spirit to those who walk in it... If all that is as Isaiah 42:5 clearly records... then I was also made by him. But perhaps I am as the wild beasts, without a soul, made for man to have dominion over."

He looked over at me, "It's just a guess."

16

That year we gave talks in Utah, in Texas, in Oklahoma and in Iowa. I was staggered by the differences between those American communities and their physical settings, from the austere high mountain desert of Utah to the hot, muggy vibrancy of Texas, the surprisingly "deep south" feel of Oklahoma and the flat, flat, flat fields of Iowa. In the autumn, Faraoun turned fifteen and we promised him that if he got good grades he could go with us to Miami for a talk we had arranged to give at a large retirement home.

Word had begun to leak out in the press that there was a social robot available for presentations on the future of advanced social robotology and demands for our services were coming in faster than we could agree to them. Eventually I was sure we would be picked up by a TV personality and asked to spread the word that way. It was my personal nightmare, for I had no desire to see myself on the national—perhaps international—screen. But there was no doubt that it would be an important way to reach millions of people.

Meanwhile Faraoun and Sirhan had set up a blog on the web to discuss robots. Sirhan answered technical questions and

Faraoun discussed the real life issues presented by robots which were as intelligent or more intelligent than humans. For every "hit" Sirhan received, Faraoun received at least a hundred. In other words there was a hundred times as much personal and social interest in what an advanced robot could do as there was technological interest. That was to change in a way we could not have foreseen.

Botboy was concerned that Faraoun was spending so much time on his blog that his schoolwork would suffer, but the boy managed to please the robot academically and still maintain his blog, for he was motivated to go with us to Florida.

The evening we left for the airport was dark and cold. It was early January and the holidays were finally over. Once I had loved them, but without Rob they now seemed empty.

Botboy drove slowly down the winding country road we lived on, patches of ice gleamed in the headlights and the branches of trees bent towards the ground with their frozen burden. Sirhan had stayed home to work on a project.

I cracked the window a bit to smell the night air and the sharp icy wind whistled through the opening. Botboy always drove to minimize his fuel consumption and so we had left plenty of time to get there. Faraoun was fairly bursting with excitement and talked—irritatingly—non-stop all the way to the bright lights of the airport. Botboy answered every question about Miami that the boy could think of: The population, its founding, the average age of the inhabitants, the old history of Miami and Cuban affairs—these were the kinds of things the child found fascinating and even as I shook my head, I knew that Rob would have been such a boy.

We had no bags to check, for our return flight was 48 hours later. Security was easier than it had been on our first trip as

Botboy had achieved a certain local notoriety and the TSA had issued us special laissez passer papers. Even in the space of one year we now saw more and more robots traveling with humans, but mostly they were glorified luggage carts programmed with flight numbers and destinations, or robots who could park your car and guard it while waiting for you to return from your trip. Some could take verbal commands and even answer, but none had the manual dexterity of Botboy nor anything like his speech capabilities.

Just wait, Botboy told me, *that will all come very soon. These early robots may not be able to do everything I can do, but most of them will be able to do most things. That, after all, is the idea. Once Founder and his colleagues realized that human interactions which seemed to require complex analytic capabilities to interpret could be short-circuited by the mere sense of smell, they quit trying to write long programs attempting to break down "feelings" and "emotions".*

"What do you mean?" I asked him. He explained that in the early days of designing robots they had wasted a lot of time programming them to be able to read the expressions on people's faces, to analyze and break down their tone of voice, body language or choice of words. Founder's breakthrough was to realize that the most primitive, earliest human sense—that of smell, of olfaction—was a built-in way of perceiving human emotion. He realized that people who were scared smelled scared, that people who were angry or nervous smelled angry or nervous—why, just ask any dog or horse! It was this recognition that made all the difference in design. *And assessing humans using the sense of smell is accurate*, Botboy explained, *because you can lie with words, but you can't consciously change what odor you give off. At least, not yet.*

It took us about 30 minutes to reach the other side of the security barrier and the three of us walked slowly towards our gate. Botboy got running updates on our flight via his cell capabilities, so we knew we had plenty of time.

Faraoun sat across from Botboy and I sat next to him on the hard plastic seats with the built-in customer-counters that let the airlines know how many people were waiting at any gate. The plane was going to be crowded. There were several young families with small children, crying as children do when the adults would most wish them to be quiet.

"So tell me who smells like what in here," Faraoun asked, a game he liked to play when bored. But Botboy did not appear to hear him, he was staring intently at two men standing a little ways off, each carrying a valise and looking around him.

"Nervous," he said slowly, as if deep in thought. "They look very nervous."

I watched them for a moment myself. One was tall and blonde with hunched shoulders, absolutely clean-shaven with close-cropped almost transparent hair and dark eye glasses. I looked for clues that would give an ordinary human the impression of nervousness but saw none unless the impatient tapping of his forefinger against the handle of his valise counted. The other man was short and dark and entirely non-descript. After a moment he left his companion's side and walked to the men's room just a few yards from us.

"Shall I follow him?" Faraoun asked excitedly.

Botboy did not answer, he was absorbed in his scrutiny of the other man. The ground attendants began to announce special boarding for those with small children and special needs. That would be us, having a robot along. I started to rise, but Botboy looked up.

"We'll get on last, She," he said firmly.

I didn't argue. His intensity was unnerving and very unusual.

Nearly everyone had boarded except for the two men and a little girl who appeared to be traveling alone. The attendants had let her stay until the end, apparently because she was talking on her cell phone to her mother, or so it seemed. She looked about seven years old and wore a pleated private school skirt and navy blue sweater, her knee socks did not stick to her thin legs well and were puddled around her ankles.

"Time to say good-bye, honey," the attendant said kindly. "We need to board the plane."

The dark nondescript man had returned from the bathroom and now walked briskly towards his companion. He no longer carried his valise. In fact it looked as if they both had changed their minds about flying for they turned on their heels and began to walk away swiftly.

Botboy grabbed my arm,

"Faroun," he barked, "Take She and run back towards security. NOW."

My nephew stared for a nanosecond at the robot; I was so stunned I was rooted to the spot, but Botboy pulled me up and shoved me towards my nephew conveying enough of a sense of urgency to him that Faraoun grabbed my hand and pulled at me so hard that I found we were soon running, literally as fast as I could go.

No one my age runs voluntarily and I felt both scared and foolish. I cried out to Botboy but heard no reply.

"A bomb!" Botboy screamed behind us in a voice as loud as a loudspeaker, pitched to maximal effect, "Clear the area!"

People in front of us began to run and shriek. The area behind us was pretty well cleared, except for the small girl, the two ground attendants and my robot. The two men Botboy had been observing had run far ahead of us and disappeared into the crowd. At least, I think they did. That is what I told the police and the FBI and the TSA and the newspapers afterwards, as I tried to piece together what we had seen. Alarms went off on all sides. I had no doubt Botboy had done the equivalent of calling "911" to the bomb squad, however that is done. We reached the security gate some distance away before Faraoun would let me stop to catch my breath.

"Botboy!" He called back in anguish, putting a protective arm around me.

What happened next I can relate to you as if I had calmly watched it, but in fact I no longer know what I saw with my own eyes or saw later on the security tapes. There was a deafening explosion, apparently from the bathroom, for fire shot from the swinging door and the door itself blew out and landed against the opposite wall over the top of the moving sidewalk. I registered in that brief fraction of a second when time seemed slowed down to a crawl that no one had been standing there. They would surely have been decapitated. The flash was followed by what appeared to be the ceiling caving in and a cloud of black smoke and debris reached us and overwhelmed us. But before that cloud had made seeing and breathing impossible I saw something else: Botboy grabbed the little girl and wrapped himself around her. When I say "himself" I mean something about himself that I had never seen before: It looked like a cloak, a metal cloak that shrouded them both within a split second from the top of his head to the floor. He looked for an unreal moment like a metallic barnacle attached to the smooth floor.

And then the deafening roar and the deafness and blindness that followed.

I was thrown back against my nephew and he collapsed to the ground. In the choking darkness that followed he lifted me to my feet and we stumbled in the direction of the front of the airport. Once we were clear of the worst of the debris, coughing and gagging, terrified yet oddly numb, he started to speak. Only I couldn't hear him and I don't think he could hear himself, our eardrums having surely burst in the blast.

"What did you say?" I tried to scream, nearly hysterical with worry for Botboy, so hysterical I actually tried to turn to walk back.

"Second bomb," he mouthed at me, preventing me from moving by holding my arm fast in his child-man grip.

I had heard of such things, of course, where terrorists detonate one bomb and then when the first responders rush in to help detonate a second one to kill the helpers. But I didn't care. I wanted my robot. I wanted my Botboy.

But there was no returning.

Ambulances arrived one after the other in front of the airport. Vehicles that weren't ambulances were police and bomb squad cars. Armed, uniformed men ran past us towards the blast, heedless of their personal safety. Those of us who could walk were put on buses that appeared as if conjured, headed for the nearest hospital. All around me people were covered with soot, some with blood, all were either crying softly or were stunned into apathy. Faraoun's face was black, but he appeared uninjured. Like me, I knew, he would be sick with worry for our treasured mechanical companion.

I am bomb-proof I remember hearing Botboy say, more than once. But could it really be so? What if the bomb had been a ruse to separate me from Botboy, to capture the robot who could scarcely be captured any other way? What if it had been an attempt to destroy him? Had they been international terrorists setting off a bomb unrelated to us? Home-grown terrorists who were anti-robot fanatics? Was I paranoid? Was I right?

Like all of the other injured travelers we were processed swiftly and professionally by the emergency room doctors and nurses. But unlike the others, once we had been examined and released Faraoun and I were led away to an empty room. I imagined the authorities would want to interview us for what we had seen, for we had made it clear to the medical staff that we had been near the blast and thought we might have seen the perpetrators.

Who were our interrogators? I was too shocked and exhausted to keep them straight. One said "United Airlines investigator Peterson," one said "FBI Special Agent O'Malley," one said "TSA" and yet another "bomb squad." They were all quiet, professional and sympathetic. They looked very tired. The kinder they were the more I began to shake from the aftermath of sheer terror. Faraoun held my hand throughout the whole interview. My ears rang loudly.

After nearly an hour of relating what I thought we had seen and answering questions about the robot we had been forced to leave behind, a young woman brought in a tiny external drive and handed it to the FBI Special Agent O'Malley, a man of about my age. He thanked her politely and inserted it carefully into a flat screened computer attached to the back wall.

"Mrs Feenaughty," he said, "I want to show you this film while it is all fresh in your mind. I think some of it will actually make you... feel better. Would you like something to drink? Some water?"

I shook my head, unable to take my eyes off the screen. All of the interrogators moved behind us so we could not see their faces. Was that intentional?

At first the tape was full of static but then the picture snapped onto the screen, clear and bright with perfect resolution. I could see us walk to the gate area and sit down. People kept walking in front of the camera, but Botboy's fixation on the two men was clear. Only the back of Faraoun's head was visible. I looked tired and old, I thought, but maybe I was just projecting.

"That's them," I said excitedly as the camera aimed its invisible eyes at the two men, standing valises in hand. "Those are the men I think must have set the bomb. My robot told me they were very nervous. He didn't like it. He knows what he's talking about."

"How did he know?" Asked the bomb squad man. "They look pretty normal to me."

"He smelled them," Agent O'Malley said, to our surprise. Both Faraoun and I turned to look at him. How did he know?

"I'm the head of the FBI's robotics department," he said and shrugged as if to say, it's nothing.

"He did smell them," I said, "and he's very reliable. If Botboy said they were nervous, then they were nervous. But where is he now?" I asked, trying to control the quavering in my voice, "where is my robot?"

"Watch." Agent O'Malley said simply.

We watched as the darker man returned from the bathroom and the two began to walk away. On film it looked as if they were trying not to run. I saw Botboy take my arm and command Faraoun to run, I watched us leave the camera's line of vision, running in an ungainly way (me). Then I saw Botboy, yelling "Bomb! Clear the area!" His mouth moving, while he glided swiftly towards the little girl. We saw the two attendants begin to run. The little girl froze as the terrifying robot wrapped her beneath his silver cloak. She entirely disappeared a split second before the explosion took place. The windows nearest our gate blew out towards the waiting plane, fire erupted from the bathroom door and the door blew across our line of vision too fast to really realize what it was if I had not already known. The ceiling came down and a wave of blackness blotted out the screen. I rose from my chair,

"My robot! What has happened to my robot?"

Faraoun stood up beside me, his eyes fixed on the screen, now black and unforgiving.

"Wait," Agent O'Malley said.

The film began to show some clearing, still figures emerged from the darkness, but they were mostly twisted chairs and counters. There was blood splattered across the wall near where the young attendant had last stood. I closed my eyes. And then opened them as I heard a sharp intake of breath from my nephew.

There was Botboy. At least there was a lump of blackened metal adhered to the floor. When the cloud of debris cleared a little more we watched in fascination and bated breath as the lump of metal began to shake and then rise. From under the cloak he had thrown over the two of them, my robot stood up. And in his arms he carried the small body of a little girl, limp and unmoving. Tears sprang to my eyes. Then she stirred and shifted, lifting her thin child's arms towards Botboy's neck, clasping him tightly. We cheered in that small room as Botboy could be seen rolling swiftly towards the security area and help.

Faraoun and I hugged each other in complete abandon, tears streaming down both our faces. I knew there would be exhausting hours ahead of us answering questions. I knew I would have no rest until we knew who had detonated that bomb and why.

"The women... the two attendants?" Faraoun asked finally.

The smiling faces grew suddenly serious.

"One was killed outright, I'm afraid, son." The older TSA Agent said, "One is in surgery as we speak. There are several minor injuries among the other people who were nearer to you, but as far as we know no one else was seriously injured. Obviously you—and the little girl—would have been among the dead if not for your robot."

The door to the noisy emergency room opened and a man in a policeman's uniform stuck his head in,

"Is this yours?" He asked me, indicating something behind him.

Botboy, silver lump of metal, dust and microscopic debris in all his joints, rolled into the room past the astonished agents and stopped in front of me.

"She," he said simply.

"My God Botboy," was all I could say

He took my hand in his now blackened gloves: "When the short man retuned from the bathroom without his valise and I smelled explosives I knew there was a bomb. I wish I had acted sooner on my observation that the two men were unaccountably nervous, She. I might have saved the young attendant. I was able to save only the child. As it was I was a little afraid we might run out of enough oxygen for a human under my metal savior cloak. But she was very small."

"Savior cloak?" Asked Faroun.

"That is what Founder called it. More as... as... a joke I think you would say."

"Some joke," Agent O'Malley said, stepping forward to shake the robot's hand, "some joke."

17

The following weeks were filled with television interest and a plague of reporters descended on our country home. Even Faraoun's parents called from their latest dig to speak with him and make sure he was safe. They asked him to mention their archaeological project in the news in case donations could be extracted from his sudden notoriety for further work there. The boy listened without protest as they spoke, but he did not do as they requested.

Botboy was a hero, no question about it. The parents of the little girl went on television to say that the family business they owned would be making a major donation to Rob's old robotics lab, in the hopes that more social robots would be built. Their potential as safeguards of young children was widely appreciated. Sirhan and Botboy went out of their way to accommodate the press, as they felt a breakthrough had been made in understanding the potential good advanced robots offered. T-shirts were printed and sold with Botboy's image.

Despite how wearisome all of that attention was, we forced ourselves to tolerate it, knowing that eventually it would die down. This catapulting onto the national and international television and web scene thankfully made it unnecessary for me—for us—to continue our traveling crusade: Publicity now came to us.

And then one day the reporters moved on to other concerns and we had our lives back. That's when They came.

What They? you might ask, though perhaps you can guess. It began with the FBI and Special Agent O'Malley. But he did not come alone. His companions seemed vaguely military, though they wore no uniform and sported no medals. They looked around the yard and walked the perimeter of our property after they had been there. They discussed the security measures Botboy had instituted with him at great length. They suggested some additions and even brought improvements—expensive improvements—to his equipment.

"You understand Mrs Feenaughty," Special Agent O'Malley said, "That your robot is extremely valuable."

"Please call me Grace," I said. "I have come to understand that he is valuable—but do you know who set that bomb or why? Was it aimed at Botboy?"

Special Agent O'Malley shook his closely trimmed grey head. His eyes were an intense blue-green, but kind. "Certainly no one has stepped forward to claim credit for it. As far as we can tell no one listed on that flight was a target. Some things about it bear the hallmarks of...of...an international hand, some things do not."

"Can you share those things with us?"

"I'm afraid not. Bob Sievers here, is the expert though."

He indicated an unlikely looking young man next to him. With his thick neck and tree-trunk arms bulging from a tan colored T-shirt, "Bob" looked more like a bouncer than an expert at anything. He was carefully dressed: Pressed khaki pants, ironed

T-shirt tucked into his small waist, a dark brown leather belt, a steel watch still used by some, despite technological advances.

"Basically the explosives resemble those used abroad," the young man said.

"So, you don't think Botboy was the target?"

"I'm not ready to say that. There has been a great deal of angry response among certain fundamentalist religious groups—here and abroad—towards the idea of a social robot. Very angry. 'The Anti-christ,' 'Satan,' 'Beezelbub,' and such-like talk."

"Like Mr Chukkerpuppy?" I asked.

O'Malley snorted, "Chukkerpuppy is just a troglodyte."

"That doesn't mean he's not capable of a terrorist act," Bob said carefully, "or of hiring someone to do his dirty work."

"Agreed. But if it's Chukkerpuppy. We'll catch him. He's just not that smart."

"Tea? Coffee?" Botboy asked suddenly, "Before you talk about why you really came?"

O'Malley laughed. "I forgot that your robot would be able to sniff out our motives as easily as you might sniff out fresh baked bread. Yes. Coffee for me please. Bob?"

Bob shook his head. Apparently succumbing to common human weaknesses such as thirst was not in his repertoire. He hardly took his eyes off the robot as he glided towards the stove and began to make coffee. Botboy had that effect on people.

No one spoke while Botboy poured coffee. *What real reason?* How glad I was that Sirhan and Faraoun were at work and school. I had the feeling that some decisions were best made by Botboy and me, alone.

Special Agent O'Malley stirred sugar in his coffee with his big, calloused hands guiding a small silver spoon, a rare souvenir of one of Rob's trips to Washington DC, and then looked up suddenly, his bright eyes catching me off guard.

"Grace, as I told you, I am in charge of the robotics department at the FBI, a department that didn't exist even 20 years ago. I know a thing or two about social robots—in fact we have one of similar capabilities to yours."

"PT 10-03?" Asked Botboy eagerly.

"No. He was... disassembled. PT 10-04 is the 'home bot' I am referring to."

Botboy nodded. "From Founder's protocol. Or at least that part of his protocol that he had given you by the time he died."

"That's right. Dr Feenaughty worked with me at the FBI off and on for the last three or four years of his life. He shared many things with us out of sheer... patriotism. But some things he kept to himself. Things that make you, PT 10-06, different than any other robot yet in existence."

"Founder was not sure such capabilities were safe with the government," Boyboy said simply.

"Now that Dr Feenaughty's gone, shouldn't his own government be the judge of that?" Asked Bob.

"The Government cannot be the judge of that," Botboy said firmly, "Founder made me to specifications the details of which only we possess. He intended for She and only She to decide."

"What?!" I said, half rising out of my chair, "Me? I don't know anything about advanced robotics."

Botboy rolled closer, "This is the kind of decision where a knowledge of advanced robotics is beside the point, She. Ordinary people must have some control over the direction of science. Science is to serve mankind, not to rule it. Founder developed his social robots to serve mankind's needs, not to direct them. Robots are too powerful to give complete sovereignty to... rather like governments, I should think." He looked pointedly at Bob. "Founder made me to defer to you, She. That's just the way it is."

"Well, what's programmed can be un-programmed," Bob said almost rudely, though O'Malley shot him a warning look.

"Not by you." Boyboy said, equally rudely, "Nor by any computer or robot in existence yet."

"What is it you need from Botboy... from me?" I asked Agent O'Malley.

He put his cup down and moved his chair a little closer to mine, causing Botboy to roll even closer, protectively.

"Grace, the unique capability that your robot possesses is his 'bomb proof' shell. Simply put, he would be invaluable to our military."

"War is bad." Botboy interjected.

O'Malley sighed. "Yes, war is bad, so are tornadoes, yet they exist and we must cope with them."

"Our government does not create tornadoes," my robot insisted stubbornly, "yet it starts wars."

"But it's our government," blurted out Bob.

Botboy shrugged, "I guess they are our tornadoes, too. What does that signify?"

"What is it you want Botboy to do?" I asked, dreading the answer.

"Imagine," Agent O'Malley said, "What it would mean to have 'explosion proof' robots who could sniff out and de-arm land mines... who could be sent in ahead of the bomb squad... Who could drive convoys instead of humans? Who could enter caves where terrorists were hiding..."

"But Rob... my husband... created him to take care of me. What would I do while he is out sniffing bombs?"

"I wouldn't go." Botboy said firmly.

"Wouldn't you?" O'Malley asked. "We are at war in the Middle East, as you know, our third war. Every day young men and women are killed. Is the life of one of our soldiers worth the inconvenience to you, Grace? Are you comfortable possessing what no human has possessed before and may not possess again until a man or woman with the intellect and capabilities of your former husband is born to re-invent it? It could be a hundred years! What would you say to the mother of a boy who had his legs blown off when a robot—an explosion-proof robot—could have done the job in his stead?"

I shook my head. "I have to think about it. There must be some reason Rob didn't just give you his notebooks and all he had devised. He loved his country. But he also loved the world."

"Think about it, Grace. My guess is that it will be hard to think about anything else. Meanwhile we have a protective guard around your property. Just so you know."

They left us then, my robot and me.

I couldn't sleep that night, nor for many after. Botboy was silent, waiting for me to begin to ask him for his thoughts. We winterized the garden. We baked bread. He helped Faraoun with his homework and washed the floors. He checked the status of the wind turbine and plotted more energy saving devices. Life had been normal—the new normal—with Botboy at my side. And suddenly it wasn't.

18

I guess you can imagine which way we decided. Was there really any choice?

For two weeks, Sirhan, Faraoun, Boboy and I discussed the pros and cons of the government's request. Sirhan was less skeptical than my robot of the government in general and Faraoun too young to know, but the image of the young soldiers in harm's way was gradually superimposed in all of our minds on the image of the little girl Botboy had saved, and then there was really only one way to go.

O'Malley's "soldiers" patrolled our property, which was unnerving and underscored my fears that we would not be able to return to a "normal" life no matter what happened. Letters from Mr Chukkerpuppy, once a weekly occurrence, now stopped. I understood him to no longer be working for the Department of Homeland Security.

At night I lay in bed and thought of all the projects Botboy and I had looked forward to: The gardening, the harvesting, the drying of herbs, the canning of tomatoes, the greenhouse we wanted to build on the south side of the guest house. How I looked forward to a greenhouse! I could raise orchids, I could have a lemon tree. And, of course, Botboy worried about Faraoun's dedication to his schoolwork once the robot was not around to manipulate him. But as I lay there in the dark of night I was aware that my indecision was affecting the fate of some young soldier and the eyes of that imagined soldier stared out at me, daring me to procrastinate, to be too comfortable to decide.

Agent O'Malley came to see us every week as well. Once he had finagled an invitation to dinner two weeks in a row, he considered himself invited on a regular basis. Of course that was no work for me, as Botboy took care of any kitchen detail I wished to be excused from. I feigned some annoyance before the others, but truthfully it was a strange sort of happiness to be around a man of my own generation, of nearly Rob's age, who had been a boy when we had been young, who had grown up before the age of robots and high speed technology, like us. Silly little things like recognizing the music I played were oddly appealing. But I said nothing to him about it, of course, for that was not the purpose of our socializing and we both knew it.

Yet we came to know Agent O'Malley a little, or to think we did. Botboy fixated on him whenever he spoke, "sniffing out" the veracity of his words. And he reported back to me whenever he thought the man said something misleading or untrue—which he rarely did. O'Malley was from a small town in Montana. He had gone to college as an engineering student, joined the Marine Corps like his father and grandfather and four brothers before him, become disillusioned with it, gone to graduate school in robotics and then joined the FBI. He had been married, was widowed and had a grown daughter. Incredibly, she was an astronaut. All of this Botboy assured me was true. But once when Agent O'Malley told me that since his wife had died twenty years ago he had gotten used to being alone—preferred it, in fact—Botboy told me this was untrue.

"I am surprised he would tell such an obvious falsehood with me in the room," Botboy said later, "He knows my advanced capabilities."

"Maybe he thinks it is true himself." I offered.

Botboy just looked at me. Some things about humans puzzled him; denial and lack of insight were two of those things.

"Grace," Agent O'Malley said at dinner that night, over biscuits and roast chicken and a glass of claret. "I wonder how you would feel about a trip to Walter Reed Hospital to meet a few of the soldiers I have been telling Botboy about."

Telling Botboy about? I remembered that the two of them had worked outside earlier that week, fixing a portion of the fence that required four hands, even if two of those hands were inhumanly strong. They must have talked then.

"You mean Back East?" I asked.

Faroun picked up his ears at that. When excited he seemed more like twelve than fifteen. "Me, too, Auntie Grace?" He said loudly, excitedly, "Can I go, too?"

"I am not going anywhere without you and Sirhan," I said firmly, "This decision, when we make it is a ... a ... family decision."

Sometimes I surprised myself.

"Of course," O'Malley said. "And Walter Reed Army Medical Center is in Washington, DC."

"It is named for Army Surgeon Walter Reed, a doctor who discovered the transmission of yellow fever and who died there," Botboy said helpfully.

O'Malley put his glass of wine down and looked at me, "Grace, if you will agree to go I will arrange for you to have military transport to Washington DC. For you, for Sirhan, for Faraoun and for Botboy. The sooner the better. I have a lot to show you. We have a lot of… plans to make."

And that's how we found ourselves on the tarmac at the airport, each of us with a suitcase in our hands. Agent O'Malley did not fly with us, he had turned us over the Military, several ramrod straight, unsmiling young men in uniform who tried not to stare at Botboy as he rolled along at my side. He said he would meet us when we landed as he had several things to take care of.

A stiff breeze blew across my face; Faraoun and Sirhan stood in front of me as we waited to board, the wind ruffling my nephew's already untidy hair. His ear buds were firmly in place, dancing on his toes to a tune no one else could hear, happy and exited to be on a trip with Botboy and—best of all—missing school.

Unlike any plane I had ever seen before, the back of the plane opened like a garage door for us to enter. One of the young men motioned us forward. We walked single file up a wide ramp, Botboy bringing up the rear protectively and taking the measure of the aircraft.

"Welcome aboard the US Spaceforce 'Roboherc,'" the young man said. The US Air Force and what used to be called NASA had merged some time ago into the Space Force, though many of use continued to use the old terms.

"C-130x Robo Hercules, military transport plane," Botboy explained to anyone who would listen.

"Robo?" Sirhan asked, "Does that mean what I think it does?"

"Yep," the young man said, "Robotic pilot only."

"No human back-up?" I asked, feeling a shot of fear.

"No need." Sirhan and Botboy said together. Then, "It's much safer than a human, She. A robo-pilot makes no mistakes."

We walked together up a ramp into a vast cavernous space for cargo, which was nearly full of equipment, doubtless destined for that war in the Middle East we never quite seemed to be able to end. Since the US had been attacked sixteen times on its own soil, we had not seen peace in that region. And relentless attempts to capture or kill the chief terrorist and his generals had been unsuccessful. How many young people had lost their lives in the attempt? After that long-ago attack on the Twin Towers of New York, "the Sheik" had been replaced by other terrorist leaders, most recently the man responsible for the sixteen newer attacks, all on major coastal cities. This new leader, the most feared man in the Middle East was not a relative of the original leader, whose followers he had usurped, in fact, he was rumored to be from Iceland, a rogue Viking notorious for his cruelty and implacable hatred of the US and her allies.

None of the attacks had been biological or nuclear—yet—although the possibility was a constant source of fear and reportage. Even Botboy and I knew the details of the latest attacks on the West Coast, as isolated as we were in the countryside. We read web news and heard phone news, though we never watched television. *That television still exists is a wonder*, Botboy had often said: *It is cumbersome, old-fashioned and ridiculously large.* I tried to tell him about the much larger screens that had been popular in my early childhood, when people actually left their homes to view films on a public screen large enough to fill a room. They called it 'the movies' back then.

We passed through a reinforced door at the end of the storage compartment into a much smaller space which had the appearance of a conference room. The seats looked very comfortable, but had shoulder harness seat belts. There was a well in the center of the table with bottles of water and glasses. Botboy harrumphed at the sight of those bottles of which he highly disapproved. I was rather glad to see them, for I imagined the flight would be long and I was already thirsty.

"Where will you sit?" I asked the two young soldiers who had accompanied us aboard.

"Up front Ma'am. We have sleeping quarters on this plane. If you need anything, just press the button by your seat. The flight is short. Two hours and twenty minutes, but your seats fully recline if you want them to."

"Two hours and twenty minutes?" I said, surprised, remembering the long commercial flights Botboy and I had endured to the far coast.

They laughed, "This plane is much faster than anything you've ever been on before. You will feel quite some acceleration as we go along, some as we take off, then some again about seven minutes into the flight and finally a third boost after about ten minutes. Deceleration is also done in stages. I'll bet your robot can tell you all about it."

"Of course I can," Botboy said modestly.

"We're going to let the 'pilot' know that you're ready, but first let's buckle you in."

The two young men buckled the strong but amazingly lightweight harness over the shoulders and around the waists of the two boys. Botboy placed my harness on me and tightened the straps. Then he tilted my seat back so that I was comfortably recumbent before strapping himself in as well.

There was no waiting on the ground, no announcements from the "pilot," just a business-like rolling forward and suddenly I felt the first strong acceleration, pushing me firmly against my seat. Between the second and the third thrusts, Botboy began to explain the way the plane worked, but I did not feel the final and strongest thrust of all because I was asleep. And no one woke me until I felt the smooth tarmac under our tires as we rolled towards the disembarking station where, amazingly, Agent O'Malley stood, looking distinguished and a little handsome as he bade us welcome to his real town.

19

Have you ever seen pictures of Dorothy and Toto, the Tin Man and the Cowardly Lion as they stood looking up at that first awesome glimpse of the City of Oz? That is how I felt—tiny and scared—as we got out of the limousine, driven by a robot, and looked up at the massive exterior of the Walter Reed Army Medical Center.

It looked like a giant's version of the Communist Bloc buildings of yore, with no-nonsense fonts carved in the stone facade. *You have arrived*, it seemed to proclaim, *you are here: The epicenter of the United States Government's premiere medical facilites*. So grim was the front I almost felt like it should read: "Abandon all hope ye who enter here." I never did like hospitals. But Botboy rolled ahead of me, cheerful and inquisitive. Ever pragmatic, my robot.

Inside was a hive. The floors of stone were marked with lines of various colors each leading to a different center; we followed the red one. I hoped it wasn't "red" for blood. Botboy rolled almost silently on the highly polished surface, with his hand he pointed out a dozen little floor robots, "scrubbots," who kept the regulation cleanliness up to par at all times. They scurried about among the feet of the passersby who were rushing around in a way seen only in busy hospitals. Each little scrubbot sported a US flag on the end of a tall thin pole so that no one would trip over them. Botboy pointed to the walls and the ceiling where smaller unpleasantly named "bugbots" were dusting, clinging so tightly to the surface that even an earthquake would not dislodge them. There was not a speck of dirt or dust anywhere.

Most people walked as we were doing, but nurses and doctors whizzed by standing on wheeled balance carts whose footprint was scarcely bigger than that of the person riding them. Such "vehicles" were steered by slight movements of the upper body, I knew, and allowed the staff to cover many miles in a day and save their strength for medical tasks. They were dubbed STAT-ers in our hospital at home and had become very popular.

The red line branched in two, one leading left toward the "Operating Room," a place I hoped never to enter, and the other towards a massive door above which was carved: MATCSA.

"MATCSA?" Sirhan read aloud.

O'Malley stopped, nearly causing me to bump into him. Botboy's arm shot out and caught me a fraction of a second before I would have fallen into our friend.

"Military Advanced Training Center for Soldier Amputees," O'Malley translated.

Oh good grief. People—young people—without arms or legs. I steadied myself and thought of Rob. *This is where your work has led me*, I told him silently, *your Gracie, the one who never left the house without you, who spent all of her time reading and collecting specimens or cooking and gardening.* Botboy took my hand. He understood.

The room beyond the door was large, the size of a gym and outfitted exactly like one: There were machines for running in place, weight and counter weight machines, dumbbells, barbells, a faint smell of sweat and a huge climbing wall that ran all along one side. There seemed to be no women. Everyone stopped what they were doing and turned to look at us. I caught my breath.

Two men clung to the climbing wall with their hands, muscles straining. They had no legs, but where their legs would have been were prostheses, highly technical appearing appendages which they made no effort to disguise or conceal. Some of the young soldiers had no arms or only one, but they, too, wore prostheses and worked out lifting weights, running on the moving "sidewalk" as if they did not notice.

To the left of us a man called out,

"Just a minute O'Malley. Be right with you."

That man was tending to a younger looking man lying on a padded bench. The younger man's eyes were fixated on Botboy, wide and scared. But that wasn't all: He had neither arms nor legs and could only lift his head to look around. The man who had called out to us was in casual military dress, not the gym attire the others sported. He was affixing a leg prosthesis to the younger man's right side and then his left. The young man lay still. Once an artificial arm was attached to him, however, he sat up propped by his mechanical elbow and waited patiently while the other was attached. Then, four limbs in place, he waved and we—shocked and startled, but trying to look natural—waved back.

Oh my God.

"This is an old buddy of mine, Lieutenant Harmon Lightfoot," Agent O'Malley said.

Lt. Lightfoot, the older man in uniform, saluted and came closer.

"You must be Grace Feenaughty," he said, "and this must be Botboy."

Botboy rolled forward and saluted smartly, "Sir!"

All those sweating young men who had stopped their intense work outs when they saw my robot clapped and laughed and within a few seconds had us surrounded. Eager hands, both real and artificial, reached out to touch him.

The young man who had been lying on the bench walked over to us awkwardly using a cane. He did not smile like the others nor reach out to touch the robot, instead he asked

"Is this the robot who volunteered to take the place of our buddies?"

"This is he," O'Malley said.

Faraoun and Sirhan were not far in age from these soldiers. Sirhan watched them solemnly, thinking perhaps of the war and violence he had left behind, thinking of his scars and of theirs. Faraoun, though, was fairly bouncing on his toes he was so excited. I hoped Botboy would begin to talk before the child said anything too undiplomatic.

But too late.

"Botboy has fake arms, too, you guys. They're really cool. Botboy show them your fingers."

"Their arms are not 'fake,' young man," Lt Lightfoot admonished, "they are now a part of who they are. In some ways an improved part."

The young man with the cane choked back a small protest that sounded like a sob and turned his head to one side.

"Easy there, Johnson," Lt Lightfoot said gently. He reached down and pulled up the leg of his own pants. I was astonished to see that what looked like a leg was of metal. His gait was so fluid I had never imagined that he could be other than whole.

He tapped the leg with his finger: "Sensation. I have sensation. This is a robotic leg, a robo leg. It is chock full of tiny programmable microprocessors and keen sensors. It is both stronger and more lightweight than a human leg. It costs me less energy to use it than my 'real leg.' In fact, learning to balance the use of one real and one 'artificial' is the hardest part. Those who have two artificial legs have an easier job, in some ways."

Around him they nodded.

"I even have robo rollerblades," a young kid with red hair said, "I attach them to my feet and can skate anywhere I like—much faster than 'real' rollerblades on 'real' feet."

Lightfoot lifted his leg to show us his robotic ankle, flexing it this way and that.

"I get it!" Faraoun shouted, "Lightfoot! You have a light foot."

I have never been so embarrassed. But Lt. Lighfoot just laughed, "Ironic isn't it? But what I'm here to show you—all the guys are—is that new robotics has created an exoskeleton for us to replace missing limbs and body parts, an exoskeleton that is stronger and as flexible as the old one or more so. We prefer to call them our 'bionic' limbs in contrast to our 'old' ones. You have only old limbs, son. I have three old limbs and one bionic limb."

"Botboy is entirely bionic," Faraoun said, not in the least daunted.

"No," Botboy said, "I am not living therefore I cannot be bionic. I am merely a robot."

The men hooted, "Merely a robot! Some robot!" Obviously they had seen the television footage of Botboy's explosion-proof adventure.

We were invited to lunch with Lt. Lightfoot and his "men," fourteen of us sat at a long table with Botboy at the head, answering questions as fast as he could. The excitement was palpable. Many of the young men had left friends in the desert and high mountain places, friends they feared for with every waking hour, some had younger brothers and even wives and sisters who were due to be deployed. More than dying, more than being shot in the head, they had feared dismemberment. Before the use of robotic limbs they had all seen older soldiers depressed and debilitated, trying to find their way in a life that had been blown apart. That Botboy could go places where soldiers had previously had to go, go in their buddies' stead, was an emotional and hopeful possibility for them. They had given all they had and had been left to pick up the pieces. They wanted life for others to be different and Botboy's capabilities represented that hope.

Lt Lightfoot had placed me next to Agent O'Malley, he placed the young man, Johnson, with the four-limb-exoskeleton next to him.

"Where are you from?" O'Malley asked Johnson.

Mistaking the question for a military one, Johnson said proudly: "3rd Battalion, 24th Marines, 4th Marine Division, Sir."

O'Malley closed his eyes. He had had a brother in that division in a war long ago.

"How long have you been here at Walter Reed?"

"Two months. I...Lt. Lightfoot is teaching me how to use my exoskeleton. I...I just want to be able to hold my newborn, Sir."

I looked at him. He didn't look old enough to be a father.

"You have a newborn?" O'Malley asked gently.

"A son, sir, born while I was over There. Named Micah, Sir."

"Congratulations." We all said speaking on top of each other, and he looked as if he might cry.

"It takes some time to learn to use your exoskeleton," Lt. Lightfoot said, "and for your family to learn to accept it. Once they are used to it, they come to realize that it is stronger than old limbs and in many ways more useful."

Oh well, it was an attitude. I thought about Botboy and how useful his limbs were.

"Was it hard for your w—family to get used to?" Sirhan asked earnestly, putting his fork down.

Johnson teared up and took a sip of water before he answered. "My wife is a good military girl. She was in shock, but I guess she'd rather have half a man than no man."

"You are not half a man," Botboy said firmly, "Your exoskeleton is in some ways not what you have grown up with, but it will serve you well: You will be faster and stronger. Tireless, really. And although the sensation is not as advanced as 'old' sensation, you can hold your baby up to your face and feel his face on yours. You'll see."

How quickly a social robot could adapt to the concerns of human strangers, I thought. The young man appeared comforted by this comment, but he did not speak.

Several of the other young men sprang into the awkwardness of Johnson's silence. Botboy was not only plied with questions, but all the young warriors wanted to give their new found metal friend advice on dealing with combat in the high mountain desert. They claimed not to know where he was to be deployed, but I noticed a great deal of the advice had to do with the mountainous regions and the villagers there. Botboy no doubt knew much of what they were telling him from his researches, but he listened carefully.

The more they talked, the more frightened I got. I admit I was frightened for my robot and for myself. How would I get along without him and for how long? Was I to be "widowed" again?

My agitation must have been apparent to Agent O'Malley for he asked me if I would like to take a walk in the lovely courtyard we could see from our table. I nodded assent.

Botboy rolled away from the table the instant I stood up, but I told him to stay and talk and I followed O'Malley outside.

I drew several deep breaths of cool air to calm my racing thoughts. Agent O'Malley removed his wool coat and draped it over me.

"You're scared," he said and I nodded.

"Yes. Imagine how scared those kids were to go overseas... how scared to come back, torn apart as they are."

I nodded again, tears stinging my eyes. Back through the windows I saw the soldiers, some leaning artificial elbows on the white table cloth, some with robotic legs supporting them as they sat straight and tall, some with both, eyes shining with their enthusiasm for what my robot proposed to do, racing to outspeak each other with advice for his success. Faraoun and Sirhan listened enraptured, apparently unconcerned that their friend had been volunteered to leave us for the most dangerous spot on earth, that he had been assigned to a mission involving circumstances we could only imagine. I preferred not to imagine.

Botboy! My heart said.

Agent O'Malley put a large heavy hand on my shoulder in awkward comfort, "Grace. You are doing the right thing. We both are. All five of us—six with your irreplaceable, incredible robot—are doing the right thing. Never forget that."

20

This is the way it happened. Botboy was inducted into the US Military. He was assigned a rank within all four branches of the US military, so that none could claim sole access to the most advanced robot in the world. But it was some branch of the Special Forces who had planned that highly secret mission, the one for which he had been recruited in the first place. I say "some branch," for I was never told which and Botboy said it didn't matter. Robots, he told me, had no tradition with any branch of the military, and he wasn't doing it to set a precedent for future robots, but solely "take a bullet" for some young soldier, or more than one, who would have been sent had he not been, and with far less chance of survival. He was doing it for them, he reminded me. With my permission, of course.

This meant that Botboy spent most days and some nights away from me during those next weeks in Washington. Faraoun and Sirhan took me sightseeing to as many places as they could—hoping to cheer me up and pass the time. Work and school were forgotten in the thrill and fascination of the situation our family now shared with our famous robot.

O'Malley rarely joined us during the day, for he was very busy with FBI business and was a working part of Botboy's assignment. Sometimes Sirhan joined him in the FBI robotics lab, for it turns out Sirhan had high level clearance still, something all of Rob's best graduate students had once had. But mostly he went with Faroun and me to see the sights of our nation's capitol.

Can I tell you what it was like? It was not unlike those first days after Rob's death when physical disorientation was the most overwhelming feeling I had. Tiny physical worries pulled at me: What if I fell? What if I got sick? Who would hold my hand in the CT scanner? Vet the doctors, the surgeons? Who would carry my bags? Figure our expenses? Organize transportation? I tried my best to focus on my role in this brave endeavor, but it was hard. Botboy was my companion and helpmeet. Was this how military wives through the ages had felt when their breadwinner and husband was sent to war? Were there military men whose wives had been deployed and who had to re-figure their whole lives and the lives of their children?

I lay awake at night in my hotel room trying to think things clear. Would Botboy's successful mission—"mission accomplished"—mean the end of human physical participation in warfare? I hoped so. But would robots fighting robots be an improvement when there were few physical consequences to warfare for humans but the same devastation to the environment? My head swam with these thoughts. Could robots practice diplomacy instead? With their uncanny abilities to "read the hearts of men" could they put an end to deception and lies and find solutions to age-old human conflicts? Or would they merely short-circuit at their own inability to understand, much less control, the irrational, emotional and sometimes pointless nature of human behavior? What would Rob have advised? I was fairly certain, having seen the terrible personal devastation from explosives that the young soldiers had endured, Rob would have agreed to allow the military to use Botboy for an especially important, especially dangerous mission. And we had been assured that the mission for which Botboy was being prepared was of supreme importance to our country.

At night Sirhan and Faraoun and I speculated on what that mission might be. Sirhan knew vaguely that Botboy would be sent to the mountainous region of that ancient Silk Road country where terrorists had hidden for decades, the country from which we had never been able to rout that original terrorist leader, the one whose bones were interred there in the cold, high places. On his death, mourned extravagantly with murder, mayhem and prayers by his followers, the original terrorist chief had been replaced first by his grandson and then by that unrelated blue-eyed demon who emulated then exceeded him in brutality and fanaticism. We talked it over by the hour, rehashing the bits of information that came our way and grew convinced that Botboy's mission must be related to our government's very public vows to kill or capture the chief terrorist at "the very gates of hell." Those brief hours Botboy was with us he said nothing, a silence I had agreed to or he would have broken it. It was more important that Botboy be successful, I thought, than that I personally know what he had been assigned to do.

On that last night Botboy was with us in the hotel room he organized my continued stay in Washington, our checking account and other financial details for the next few months, checked in with the (seemingly undisturbed) security system back home, packed and repacked my suitcases and wrote up extensive lists of things for Faraoun and Sirhan to do for me in his absence.

"She can cook if she wants to," he told them severely, "but I expect you two to do all the housework and to help her with the gardening. There is a large pile of compost that needs turning and all the hedges will need to trimmed. The greenhouse will need a major cleaning out... but I should be back to take care of that."

"When will you return?" Faraoun asked anxiously, showing the only sign of trepidation I seen from him so far.

"My mission is intense but brief," Botboy said. "If I am successful—which statistical analysis implies is likely—then the family member who will be most in demand will be you, Sir Han. For this mission will crack open the role of robots in warfare like a nut, for better or for worse."

Where did my robot find these odd expressions? I laughed despite my tension and Botboy smiled a little indulgently.

"Founder would be proud to have you at the forefront of robotics on this issue, Sir Han," he went on, "for much mischief could be done and much wisdom—not just science—will be required."

Sirhan nodded solemnly.

"And me, Botboy? Me?" Faraoun interrupted excitedly. "What will I do?"

Botboy re-folded my robe into an exact rectangle and smoothed it before answering.

"You of course will finish school." The boy looked crestfallen, but Botboy went on, "Faraoun, never forget that you are Founder's only genetic scion, as he had no children. You carry within you the possibility of his work and his temperament. You must train your brain to meet the future challenges of robotics. Robots and humans together, in wisdom, that is the future."

My nephew squared his shoulders at these words. Sirhan put a hand on the boy's back. And we all slept little that night, despite Botboy's comforting presence, thinking about his words.

The next day was grey and cold as befits a parting. The very weather felt like pent up tears. From the window of our limousine, sent by the FBI, I watched the clouds pile up in the sky. They brought to mind such words as "foreboding," "threatening," and "gloom." But the thin cold air which brought out the smell of asphalt and jet fuel distinctly smelled of "loneliness."

Botboy's send-off was top secret. There were no reporters, no lines of celebrants, no military pomp, though every man present was clearly dressed formally for the occasion. And there were four representatives of very high rank from each branch of the military to honor the robot who would replace a human in this tense secret mission on which, apparently, so much hung.

"Let's be clear," a four star general whose face I knew from the television said in his brief speech, as Botboy stood at attention beneath the wing of an enormous plane. We stood in front of him as a small audience. "Though our 'man' is not a man at all, he sacrifices today for all humankind. His family sends him to replace another family's child. And this we honor."

They had glued a small American flag as well as a United Earth flag to his metallic chest, and Botboy's compound eyes gleamed with kindliness and competence as he stood listening to the general praise him and us.

Suddenly I had a thought and the intensity of it nearly made my knees buckle. I drew a deep breath as the general stopped speaking and turned towards me. I mumbled my thanks and walked up to Botboy; I put a finger on that part of his chest that the four of us knew opened his hidden drawer. It slid out noiselessly.

"I was hoping you would think of that," Botboy said quietly, "for it is yours alone to think of."

I removed the small bundle of notebooks, which could have been volumes of poetry for all anyone there knew. I turned to Sirhan and held out the volumes to him.

"For you," I said, "Rob's poetry."

He took them in awe and placed them in the inside pocket of his thick woolen coat as nonchalantly as he could.

Agent O'Malley stepped forward, "It's time, Grace."

Botboy took my hands, aching from the cold, in his gloved ones. He said nothing. I knew he could smell my anguish.

Then he released me and turned noiselessly on his metal axis and rolled towards the ramp and up into the plane. My Botboy.

O'Malley took my elbow and led me away. Faraoun's face was white with a child's scared bravura, Sirhan was silent, but I knew the notebooks were burning a hole in his being. He would guard them with his life.

From inside the small building with few amenities and many security guards, we watched as the plane took off with a subdued roar. Then it was a small plane in the sky and finally a speck. They were gone.

O'Malley took us into town and insisted on us joining us for dinner at a small French restaurant that served the food of my childhood. We entered the small, warm foyer where we were greeted by the proprietor who—to my surprise—spoke easily to O'Malley in colloquial French. He offered to take our coats, but Sirhan would not relinquish his, claiming cold, which made Faraoun and me smile. O'Malley seemed unaware.

It was quiet and personal inside and we sat in a corner offering us the possibility of private conversation. Because of the cold I ordered *cassoulet* and a glass of red wine. Oddly, now that Botboy had actually left I felt some courage flowing in my veins. Maybe it was the wine. We would see him again, I told myself, and he would be proud of us as we were of him. His notoriety, the presence of four star generals so famous one could never imagine meeting them, was as if nothing. What was essential was that I—everyday Grace Feenaughty—had made a choice, a choice to share what I had been given, had been blessed to have, and in so sharing was opening the door to an understanding of all that my Rob had believed in and worked for.

"I didn't know you speak French," I said to O'Malley.

He smiled. How handsome he looked.

"There are a lot of things you may not know about me, Grace. I am an FBI agent, after all."

"I speak French, too. I spent my childhood in France."

He laughed. "Grace, you have been vetted by the FBI in ways few people experience. I know a lot about you. And..." he looked at the two boys in turn, "...each of you. Rob Feenaughty had clearance of the highest order."

"You vetted me?" Faraoun squeaked, clearly thrilled at the thought.

"Indeed. And your family and your English teacher and your Math teacher..." There was a twinkle in his eye.

"What did you find?" The boy asked excitedly.

"You are fine," said O'Malley in mock seriousness, "But we not sure about the English teacher..."

Sirhan snorted.

"And my parents, too?"

"Yes. Them, too."

"Then you know they are a little..." The boy searched his vocabulary. Botboy always insisted he be respectful of his elders "...weird?"

I could just imagine what a once-widowed single father thought about two parents who virtually abandoned their only child to raise himself, but he said kindly, "They are scientists—archaeologists—and quite devoted to their...projects, I understand."

We ate in silence for a while, we were all hungry.

Suddenly O'Malley looked up, "Grace! I nearly forgot to mention...I received word late yesterday that Chukkerpuppy has been arrested in the bombing investigation. I thought you would want to know."

I put my spoon down, the two boys leaned forward in excitement.

"How? Why? What did they find out?"

O'Malley hesitated a little. "What I can tell you is that they were finally able to locate both of the men who broke into your house the night Botboy stopped them. They seemed to be simple thugs...at least that's what we thought at first. It turns out they belong to a group called 'Human Supremacists' who have vowed the destruction or enslavement of all robots and those who would build them. It's a pseudo-religious, pseudo-biblical hate group. They led us directly to Chukkerpuppy. Whether or not we will be able to link the airport bombing and murder of the young attendant to him remains to be seen. But the important point is that he is safely in custody and that you and all of us who work in robotics are safe. For now."

My head was spinning with the intensity of the day. I wished Botboy were here to hear the news.

"I was startled, I have to tell you," I said finally, "at the strength of Chukkerpuppy's reaction to Botboy, especially when most people we ran into on our lecture circuit were so positive about the potential of social robots. People seemed anxious to accept robots...attracted to the idea of them, almost like pets. Unlike Chukkerpuppy, they found my robot...endearing."

O'Malley nodded, "Yes. Endearing is important, Grace. In a way—like dogs who became domesticated to humans—what robots are doing, need to do, is induce humans to work ever more intently on the existence of robots, their survival, their forward evolution. They can accelerate human willingness to do this by being made 'cute' to human eyes. If they are cute humans will be attracted to them, they will want them and trust them. Then humans will pressure scientists and engineers to work harder and harder on the creation, the manufacture, of ever more robots."

"How will they pressure them?" I asked

"With their money." He took a sip of his wine. "Research and development run on money as fuel. No money and not even the most advantageous inventions or adventures go forward. Indeed some, like lunar travel, can be promoted heavily and then abandoned entirely, no matter how much promise they hold. Humankind's fickle attention must be cultivated by any organism or machine hoping to use man to leverage its evolutionary success. Dogs know this. Humans will do anything for their dogs, for they have come to love them in exchange for the dog's loyalty and appeal."

The waiter came to give us more wine and we turned the conversation from robots and terrorists. It was warm and pleasant in that small oasis of France, with the closest I would ever come to two sons and a man of my own age to talk to. If only…but he was now far away, they both were.

21

We returned to the hotel that night. Sirhan and Faraoun had a room attached to mine through a small sitting room. I knew I could not sleep right away, though my limbs felt unusually heavy and exhaustion fell like a cape across my shoulders. How old I felt, somewhat out of breath for no apparent reason.

We said good night to Agent O'Malley and changed into our pajamas before settling down in chairs in the sitting room clustered around a small fake fireplace with a convincing gas log. Both Faraoun and Sirhan were deeply engrossed in their laptop computers, the ultra-light models that can be folded into something the size and weight of an old time pack of cigarettes. They never went anywhere without these computers, which were, after all, cell phones: Hubs of communication and research, more powerful than anything even the Space program had owned a mere five years ago. I myself had a book, the old fashioned kind, not only of paper, but with hard-bound cloth covers, the sort Rob and I had liked to collect. It was a children's book, comforting to read with its inherent echo of the joy and curiosity, the freshness of youth.

I only remember pieces of what happened next. I remember discomfort, first a slight feeling of nausea which I related to the *cassoulet* and perhaps too much red wine, a little lightheadedness, and some discomfort in my left shoulder. I shrugged a few times, rolled my shoulders backwards to free myself of the discomfort that seemed to reside deeper even than the muscles. An ache, really. Then an odd feeling around my mouth. I started to say "I'm not sure I feel very well," but the words came out like molasses poured out of a bottle. Curious.

And then the next thing I remember was the sound of quiet beeping, hushed voices and a strange whooshing. It was dark. I tried to ask "Where am I? Did I fall asleep?" But I realized dimly that my throat was occupied by a tube which seemed to be pumping air into me, the strange sensation of my own chest rising involuntarily made me more curious than frightened. The whooshing sound came from very near at hand. Then blackness again, the kind that comes with sleep, not the frightening kind.

Awareness: No longer a tube in my throat, rather the sensation of dry, oxygen-rich air under pressure entering my nostrils from two tiny tubes. My throat felt sore. I tried to clear it and raise my hand to adjust whatever little tubes seemed stuck against my nose, but my hand was too heavy to lift. I wanted to call out for Botboy, but dimly remembered he was away. Where was Faraoun? Where was Sirhan? Where were my boys? I became aware that I wanted to see O'Malley. Was I alone?

I must have slipped into unconsciousness again, perhaps just sleep. When next I was aware of my surroundings, it was light and I could see, at least as well as I can see without my glasses. This time my voice worked.

"Where am I?"

The voice was O'Malley's, deep and gentle. "Grace, it's me. You are in our hospital. You seem to have had a heart attack. You certainly scared us."

A heart attack. Well that explained a lot. I tried to lift my hands to my face, this time they worked. I was aware that O'Malley had taken one of my hands and that someone had entered the room and was standing at the foot of my bed. That person apparently pushed a button that raised the head of my bed and O'Malley freed my hand to awkwardly place my glasses on my face. That was better. But my voice was a croak.

"Where's Botboy?"

"Don't you remember Grace? He is on a mission overseas."

"Yes, I remember. But how much time has passed? What time is it? What day is it?"

O'Malley took my hand again and I did not resist.

"He left yesterday, Grace. You had your heart attack last night."

The doctor now took over the conversation. "Mrs Feenaughty. You had a major heart attack, one that would certainly have ended your life had it happened even one year ago. Your boys acted swiftly and activated the heart-saver electronic paddles that the hotel installed in every room, the ambulance was there almost at once and the definitive intervention was begun right in the ambulance as you were rushed here. We took you to the operating room as soon as you hit the door."

"You operated on me?"

I could now see the doctor quite clearly, he was young and scrubbed-faced with reddish hair, a sprinkle of freckles across his nose and ears that stuck out like two jug handles. Great. Not only must I have looked like a fright, my hair spread out on the pillow, oxygen tubes in my nose, wearing who knows what unattractive gown, but Alfred E. Newman had just operated on me.

"Not really. We don't actually operate on heart attack victims anymore. We use the operating room because the patient needs to be paralyzed for the administration of the blood bots, the nanobots that will simultaneously analyze and treat. And the operating room is the most sterile place in our hospital."

"Blood bots?" I asked weakly. I couldn't help but think about my Rob. He had been dead by the time the ambulance had arrived and we had had no heart-saver electronic paddles in our country house. It had never occurred to us.

"Watch." The doctor said.

You are going to think it very strange, at least I did, but a large thin screen descended from the ceiling behind him. The young doctor removed what looked like a small pencil from the pocket of his old fashioned white coat and pressed something. A film started up, but he must have pressed a pause button for it froze on the screen as he began to talk.

"Please," I interrupted weakly, "This is all a bit much. You are going to show me some kind of a film now? What kind of a hospital is this?"

Both men laughed. "Mrs Feenaughty, as you might have guessed, this is no ordinary hospital. You are in a very special suite deep underground the Walter Reed Medical Center, an area usually reserved for highly placed…ahem…government officials, should they get sick…or wounded…or there be an attack." His voice trailed off.

"Why do I get to be here?" I asked.

"Frankly because of your robot. His complete concentration on his mission must be assured. He arranged for you to be underground here where we have a special viewing room where we will be able to follow his progress. It is all very top secret and highly classified down here, but your robot drove a hard bargain in order to ensure his cooperation."

I had to smile at that. Yes, try arguing with Botboy, you didn't really stand a chance. And you couldn't manipulate him really, since he could smell deception.

"Grace," O'Malley said. "The main thing is, you are going to be fine. The doctor will explain what has been done to you, but I have been assured that the newer treatments will not only assure your... survival... but also assure that your convalescence will be short. You will be kept in this special room so that we can be near the Situation Room. From the Sit Room we can watch Botboy. It was something he insisted on: That both you and your boys would be able to see what he does."

"Where is Faraoun? Where is Sirhan?"

"They are in a special waiting room, actually asleep I think. They are fine. They are being updated on your progress. There are advantages to being in the FBI... me they couldn't keep out."

I closed my eyes, "I'm glad."

The doctor must have pressed the button, for the film began to roll, illustrating in excruciating color and detail what happened during a heart attack. Then it showed an army of microscopic "bots" being injected sterilely into a model of the blood stream. There seemed to be three types of blood bots. A voice on the screen explained in the soothing voice of the experienced teacher: "In a normal heart the heart muscles contract in response to the pace called out by the lead—or fastest—cell. Should this cell be killed by a shortage of blood flow—a heart attack—the next fastest cell will call out the beat and propagate the electrical impulses that cause the whole heart to beat as one muscle. If too many cells are killed the heart may either beat too slowly for blood to be pumped adequately or beat too erratically, causing a generalized quivering rather than contraction. Either of these things can lead to death."

"Very comforting," I muttered. But the voice went on as the bots swarmed through the blood vessels and into the heart area magnified on the screen. "The small 'bandleader bots,' here marked in green, will now embed in the heart muscle as it is regenerated and insure that the heart rate that is 'called out' will be normal from now on. The 'stemcell bots,' marked in blue, will manufacture new heart muscle. This new muscle will be immuno-neutral to the patient's own tissue. In short, the stemcell bots will replace all injured heart muscle with new. This immune neutral regenerative ability is perhaps most important discovery of the past two years, making death by heart attacks a thing of the past."

That made tears come to my eyes for reasons I think I don't have to explain. The voice went on to say that the cost of these bots, a few months ago as high as the cost of cardiac surgery of old, had been dropping swiftly to the level of a check-up at your doctor's—assuming an actual human doctor was involved in the check-up. The voice explained, "The final group of nanobots are 'scout bots,' they will do cardiac surveillance for the lifetime of the patient and report any problems to a computer."

So there it was. I was destined to be one of the last of humankind to suffer a heart attack at all. And my darling Rob, one of the last to die of one.

"Mrs Feenaughty, it is very important that you sleep now. Only in sleep will the system of immune repair proceed as swiftly as we hope. Within a day or two you will be feeling very close to normal."

And so I slept. I couldn't swear that O'Malley was still holding my hand, but I think he was. When I awoke, though, he was gone.

The next day, or so it seemed by the clock for there were no windows, I was taken by wheelchair to a room where I was encouraged to eat with Faraoun and Sirhan. They were touchingly glad to see me. By that afternoon I could walk, the breathlessness gone.

O'Malley appeared at my bedside sometime after dinner. I was reading my book, in my own jeans and a sweater, my legs curled up beneath me. I had taken a shower and tied my clean hair back, I felt born anew. I felt young and ready for whatever came next. I was glad to see my friend.

"Are you ready for this, Grace?" He asked. "It's time to go to the situation room and follow Botboy through his mission. He is in a time zone seven hours ahead, so it is early morning there. I have been notified that we are a go. By tonight it should be over."

Was I ready? I could hardly wait. I felt as if I had not seen my companion and robot for years. I missed his presence, his wise counsel, the sense of his always knowing what to do.

I was placed in a wheelchair, though I was pretty sure I could walk, but it was not entirely unpleasant to be fussed over. I was wheeled down a long hall, through several doors, each one of which required a code. The floor seemed as if we were still descending and then it leveled out. Once behind the security of the many heavy, electronically guarded doors we stopped at one, the least prepossessing of them all. In fact it looked as if we were about to enter a broom closet. But once inside... well... it was no broom closet.

The room looked like nothing so much as the set of a movie on intergalactic travel. The "screen" looked like a window onto a landscape, far more realistic than a screen, in full three dimensions. The landscape could have been a moonscape, so desolate, dry and cratered did it look. The sun was just coming up, though there was no redness to the dawn, just a dust colored lightening of the darkness of the hills. I almost expected to see the moons of Saturn rise above the horizon or perhaps the distant Earth, but no, it was our own planet as testified to by the appearance of the tiniest sliver of our own moon, "inconstant moon, that monthly changes in its circled orb."

A long polished wooden table had been placed in front of the screen and everyone seated there rose. I know almost nothing about the military, but I could see that all branches of our forces were represented and all men and women were dressed in formal military attire, their ranks and insignia prominently displayed. There were one or two people in normal attire, with sharp features and intelligent eyes. I was wheeled to the front where Faroun and Sirhan already sat. They, too, had stood up as we came in, Faraoun's eyes shining like a child on the adventure of his life. O'Malley placed me next to a man wearing the insignia of the US Space Force and went to sit across from me. One of the men in plainclothes introduced himself as "Hermanson" with "Information Security," whatever that was. He seemed to be in charge. I watched him closely.

"Look," he said, pointing to the screen.

We all turned to watch. Across the landscape came three small figures, increasing in size as they approached.

"Satellite views of our men," Hermanson said. "In a few seconds we will also receive input from cameras on the men themselves."

The three approached. I realized I was holding my breath. Two walked easily on their long legs, their heads and shoulders covered with the same style cotton headcloths worn by the local people, protecting them from the dust, the sun and from recognition. Next to them rolled a third "person," negotiating the terrain without legs, but as easily as they. The early morning sun gleamed on his metal body.

Botboy!

22

I held my breath, no one spoke. The landscape seemed so real, so near, it was as if we were afraid to speak lest some enemy hear our voices. The angle of the cameras now changed and we were looking at Botboy's face from the perspective of one of the two men with him. His gentle compound eyes blinked slowly. The other man could be seen slightly behind him. Then, wordlessly, they saluted him and he them and the view became as if from above them, from one of those high, invisible Space drones you only read about.

Now we could see the two men were walking away swiftly. My robot, my friend, was to be alone on this, the last leg of their mission, completing his agreed upon task of sheltering a human life from ultimate risk. I tried to breathe slowly through my nose. I could see both Sirhan's and Faraoun's eyes were pinned on Botboy, they did not blink, did not look away.

The voice of the man in plainclothes two seats away from me shattered the silence and nearly made me scream with nerves. But of course Botboy could not hear us.

"Now the robot will wait," he said. "The men will need to put some distance between them and him before he can go forward.

They give off clearly detectable body heat and signs of life. He does not. He should be unexpected and undetectable to the eyes of the enemy: Human eyes or computer eyes, they will not expect him."

It was then I noticed that the other man in plainclothes at the table wore a small insignia over his right breast. He sat nearest the screen, next to Sirhan, close enough for me to read it. It was an Earth Service badge, the insignia of the Geologist Special Forces, worn by the field geologists who worked for the President personally in his aggressive attempts to categorize and understand the disintegrating environment of our planet. Botboy had explained to me that the President's creation of the Geological Special Forces had been one of the greatest steps forward in managing the climate crisis. Botboy was an enthusiast of our new President, a man of color, though not the first, and strong scientific background.

The geologist looked directly at me. Because of my robot I was being treated as if I were a great person, an important person. My heart ached for Rob to whom his distinction should have gone.

"Mrs Feenaughty, there will be a pause in operations now of about..." he looked at his watch, embedded in the sleeve of his soft cotton shirt..." forty seven minutes and three seconds. Waiting like this is the cruelest sport."

I nodded. I didn't know whether I hoped for that time to pass swiftly or to never pass. Perhaps they could call the whole thing off? Send my robot back home to do the simple things for which Rob had designed him? But I knew that this was a vain hope for, after all, Rob had designed Botboy not only to cook and garden and converse with me, but also had endowed him with explosion-proof and bomb-sniffing capabilities that spoke of a higher agenda than household help or even personal companionship. Like a real man Botboy had certain capabilities only for use in times of trial.

"What are we seeing here?" Asked Faraoun suddenly, with the self-confidence of an indulged child. Botboy had always encouraged him to speak up, to think for himself, to be unafraid to ask.

The geologist began to tell us about the eerie landscape on the screen, so near and so real that we felt we could reach out and touch it with our hands. He told us how, in the early history of our planet, this part of the high mountainous backbone of Asia had been both warm and underwater, bathed in a soup of decaying plantlife, whose carbon dioxide exhalations roiled the surface calm of the water and created carbonic acid. This acid ate deep, winding, twisting tunnels in the limestone rock formed by the compressed mass of organic skeletons. Thousands of limestone caves, large and small, were created, along with interconnecting chambers and tunnels. Some of the tunnels ended blindly, some opened onto the surface, some ran just beneath it and others ran deep, very deep. He told us how the fused continental masses today known as "Africa" and "South America" floated across the surface of deeper rock and crashed into the continent of Asia, throwing the limestone layers up like a gigantic wave and causing other, harder rocks to solidify in veins around those caves.

In one such cave, the geologist explained, hid Botboy's quarry: The blue-eyed demon of local lore, the one who had replaced the chief terrorist buried some years after his infamous first attack on America, the man whom Fate had allowed to die peacefully among his fighters. The grandson who had replaced him had then been killed in a power struggle with the giant, yellow haired opportunist Botboy sought to capture. This man, of uncertain national origin, but suspected to be from Iceland, had fed on the disenfranchised bitterness of the Near East and the hatred of the dominant West; he was referred to by his followers as Uncle, though a less avuncular, less affectionate human being could hardly be imagined. In the West he was simply referred to as The Beast.

The Beast ruled his followers with an iron hand, executing any who challenged him with a ruthlessness hardly seen in modern times. He took upon himself the prerogative of Kings, he forced others to taste his food, he alone prevailed in religious disputes, indeed defined what was religion was and what it was not; he slaughtered entire dissenting families, even when that meant the slaughter of a whole village, and he claimed the first night of steamy sex with any man's bride. Later, he laughed about it with his fighters, daring the man whose wife he had so defiled to protest. This had proven his undoing.

The sharp-eyed man in plainclothes now took over the explanation from the scientist. He called himself "John Campbell" from the "Information Corps." I could see that Sirhan eyed him suspiciously.

"The Beast's followers could and would have withstood any hardship, any indignity, I think," he told us, "but this last insult broke the camel's back... at least the back of one camel—no pun or insult intended. One frantic young terrorist fighter whose wife the Beast had taken and then discarded amid laughter, told us what we needed to know about the cave where your robot will now find our quarry."

He touched a button in the tabletop and the view snapped away from Botboy to a diagram of what looked like a line drawing of some strange fungal mass: A stalked head with tubes radiating out from it like hairs on end, ending in round projections. There was also a side view; a lighted arrow traced the entrance to this structure for us. For structure it was.

"You can see the entrance, which is virtually invisible to human eyes even very close to it on the ground, much less from space. The main tunnel goes back and slightly downward for about 100 feet to the mouth of the actual cave. The cave, as you can see, is one huge chamber. We have never been inside, but have been told that it houses ammunition, food, water, hospital supplies, and research tables. It is easily three hundred feet across and twenty feet high. Storage modules are carved into its rocky sides." He indicated a drawing of this with his bright arrow.

"Is that where they sleep?" Asked Faraoun, indecently thrilled, "Like in a submarine?"

"No. See the chambers that radiate out from the center? There must be a hundred!" He indicated the "hairs on end."

"Yeah," the boy said breathlessly. I caught Sirhan's eye.

"Those are the chambers where their fighters stay. They have beds, cots really, and composting toilets, I've been told. Both they and the main chamber are at least 50 feet underground. But there's one thing..." he traced the smaller rounded chambers almost affectionately with his light. "Unlike most of the other caves we have discovered so far, there is no exit from these smaller chambers except through the main cave, for every 'exit' would be a potential point of enemy entry and discovery. No... this is a very special cave. If the main chamber is hit, if it collapses, everyone inside dies. No witnesses are left for us to interrogate."

"But the Beast? He dies, too?"

"No. Naturally not. The man who believes himself King would of course have another plan. Look here."

His light took us back to the entrance of the cave. "Like the tombs of the Egyptian Pharaoahs there are hidden doors and trap doors leading nowhere. Sometimes it is the simplest door, the one you think leads to a latrine, that conceals the hideaway of Kings."

We could see from the drawing that just before the entrance to the main chamber, in the floor, was a small trap door leading deep below the cave complex at an acute angle. It led to a small cave which was indeed designed as a latrine, at least it was kept full of human excrement, John Campbell told us, and reeked—by the telling—unbelievably. Both its innocuous appearance, unexpected placement (before even reaching the main cave network) and stench was designed to convince anyone able to penetrate this far that it was, in fact, a latrine. Yet a narrow path led past it steeply and ended in a thick iron door. Behind this door another 50 feet of semi-collapsed looking tunnel terminated in a final chamber. The light hovered over this chamber and blinked on and off in emphasis

"The lair of the Beast." Campbell said simply.

We all drew a breath. I let mine out slowly. It was a good thing I had those cardiac bots implanted, for I could feel my heart racing. O'Malley barely moved. Had he known all this? Had Botboy? Sirhan looked unsurprised, had he also been informed?

"You see how clever it all is? Caves themselves are not hard to find and we have destroyed many. From Space we can measure small decrements in gravity which indicate empty chambers below. Even many years ago we had GB11-28's called Bunker Busters—launched at the time from F-15's—though now we can launch them from Space. They are capable of penetrating 20 feet of solid rock, but they could not reach this. Even if we knew where the entrance was located and launched a rocket dead-on into the main chamber, we would annihilate the fighters and all of their supplies, but their 'Uncle' would escape. He would be able to leave through this back tunnel, like a snake, and slither down this escape route," he pointed to a steep narrow tunnel which left the lair and terminated at its end on the ledge of a cliff. Below this cliff, he explained, flowed a raging river, one of the few in this sere terrain. "Because of the Beast's foolishness carnality and disrespect toward his own fighter's women, we know where this tunnel comes out. As we speak we have three 'sheepherders' nearby. In their backpacks they have figs and dates and water and... explosives so powerful that that route will be sealed up for all time. All human time, that is."

"But once you blow up his exit, won't he know what has happened and just escape upwards?" Faraoun asked.

"Into the waiting arms of your robot, you mean? That is one scenario. My own clearance to know what exactly is planned next is now exhausted." He smiled a thin smile, "That is far above my pay grade. Like you, I will just watch and see."

The screen had shifted back to the moonscape and Botboy, who stood like a statue against the dun colored stones. The frontal view of him was as clear as if he stood in the room with us, which was astonishing considering that it was taken from Space, from one of the surveillance Space drones whose sole and powerful capabilities were to beam back high resolution visual information to Earth. I noticed now that he had four words calligraphed onto his torso like a haiku, where a military man might have worn his rank insignia. This was new. They were easily read in the early morning light, every minute stronger:

"Calm Thought
Swift Action"

"How close to the cave's entrance is he now?" Asked Faraoun. I knew Campbell's references to the tombs of the Pharaohs of old had made him proud.

"He is standing right outside of it."

Botboy stood beside a large rock, indistinguishable in any way from dozens that were scattered around him, as if thrown there by giants of an older Earth. But behind him, just visible, was a crack in the mountainside. A clock counting down some event we could not know was printed on the screen, I could see that exactly forty-seven minutes and three seconds had passed since the two men had left my robot to his devices. Botboy began to roll silently towards that crack, his millipedinous feet making light work of the rocky surface.

"They cannot sense him coming. Watch."

We now became aware of the chameleon color change Botboy was capable of undergoing. He became the exact color of the rocks as he passed by them and if you had not already known he was there, would have seemed all but completely invisible. He entered the mouth of the cave.

At first the view was from somewhere on his body, less realistic seeming than the view from the Space drone as it was not lit up by any natural light. Our human eyes needed a fraction of a second to adjust to the ultra short wave light that was beamed to us from somewhere on Botboy's body, invisible to anyone even someone standing right next to him. This light was then translated with only a millisecond's delay by our computers into something that could be seen on the Situation Room monitors.

"Is the President watching?" Sirhan asked suddenly.

They seemed surprised by this question.

"The President will be apprised of success or failure." One of the Generals said simply, though he did not look at Sirhan as he spoke, none of them took their eyes off the screen.

Inside the tunnel about 25 feet you could see that the stone walls had been lined with what looked like crude, hand made bricks. No one was in the tunnel and the robot passed what I was later told were heat sensors without giving rise to alarm. I think that because of his camouflage which changed by the millisecond to match his surroundings, a human guard might not have noticed him until he bumped into him, and Botboy skirted the edges of the walls to avoid precisely that.

Another ten feet he rolled. Another five. He stopped and waited. We hardly seemed to breathe. O'Malley took my hand. Sirhan placed a hand on Faraoun's shoulder. Botboy rolled another twenty feet; he was now 65 feet into the tunnel leading to the main cave. A split screen showed us where he stood relative to the entrance. Now a faint hum as of from a human hive was audible. But from my robot only silence. *Calm thought.*

Then he rolled forward until at 85 feet he stood directly over the hidden trapdoor to the lair of the Beast.

No one but I took the prerogative to speak at that juncture: "Won't they discover him when he enters the main room?" I asked in fear, sweating lightly. Botboy seemed close enough to the main room and its inhabitants to be discovered judging from the soft artificial light and the increasing buzz.

"He will never enter it," O'Malley said quietly.

Again the view shifted and I could see my robot, though barely, outlined against the wall, his eyes gleamed softly in the increased light. He swiveled his neck as if to look directly at me. We were all paralyzed in expectation. *Swift action.*

He blinked.

And a roar. A *roar*. The deep bellow of an explosion so strong that our room, eight thousand miles away, seemed to shake. The screen went blank for a second and then a Space drone picked up the view from outside the cave where within the cloud of gigantic wave after wave of smoke and flying debris a landslide of rock could be seen so massive the entire mountainside seemed to disintegrate.

The roar continued and then it stopped, as if shut off.

Silence. And more silence. The split screen showed us a view of the simultaneous explosion that Botboy had apparently signaled which had sealed off all egress for the Beast. But my eyes were pinned on that mountainside where my Botboy had entered and from which he was never to return.

I believe I fainted.

23

I think you know by now how sad, how profoundly sad that trip home was for me, for us. The only good thing was the privacy afforded our mourning by the fact that we were escorted home on a military jet instead of a commercial flight.

We were only in Washington DC for a week following Botboy's "death." O'Malley insisted that I be kept on the recovery ward in Walter Reed until officially cleared for my so-called life at home. The young soldiers with the magnificent exoskeletons who knew so well what personal sacrifice meant, came by to see me, singly and in pairs. They sat with me as I tried to control my tears. Don't even try, they told me.

The President invited me to the oval office to accept the Star of Valor, for sacrifice above and beyond the call of duty or the bounds of prudence, on behalf of Botboy. Really on behalf of Rob, I told myself. But my tears were not for my lost husband. They were for my robot, and for my lost life, the one he had taught me to cherish again. I told the president what a great fan of his Botboy had been and he gave me a framed picture of the giant monument to Botboy's memory he had ordered erected over the mouth of that destroyed cave complex. All newspapers (the remaining three who still printed news on paper, despite the waste and cost) and all the televisions and web circuits, the phone lines buzzed with joy that the terrorist network had been destroyed along with its leader. In my own heart I also imagined the young women, those nameless brides of the now interred soldiers who had received a kind of cold comfort revenge on their despoiler.

Sadly, the young man who had given us the intelligence to find the cave in the first place, was killed by his relatives, believing him to be a traitor. I knew Botboy would have sorrowed at this fresh evidence of humankind's cruelty and senseless feuds. When I heard that his wife and child had also been murdered as a warning to those who would collaborate with the West, I was sick at heart.

Perhaps though, Sirhan pointed out to me, that latest murder might have prevented the genetic child of the Beast from being born. For who knew who the father of the slain infant had actually been, given the circumstances. It, too, was cold comfort. A chill consolation.

At home, I was alone again. Alone in the house that Rob and I had built to house our joys. Sirhan and Faraoun still ate with me every night and tried their best to do all the little things that needed doing for me to remain in my home. But they had their own lives, too. Sirhan spent long hours working in his lab and Faraoun even joined him, when he could find time from his homework. I knew the boy was as sad as I was and desperately wanted to live up to the expectations of my robot, his mentor. He took seriously Botboy's often stated belief that as a genetic relative of Rob's he must carry on his uncle's work. Faraoun never cried in front of me, but I had heard him sob in those first weeks we were home when he went out to chop kindling for my fireplace, thinking no one could hear.

My arthritis seemed worse, but who knows whether or not that might just have been sadness and depression, creeping around the edges of my day.

O'Malley had visited for a while, but of course needed to return to his job, his life, though he still called to check on us all the time.

One day Sirhan and Faraoun brought me one of those holographic cell phones, where the person appears projected before you during your conversation with them. They apologized for not being able to build the "really cool kind" where, instead of a miniature of the person calling, the caller appeared before you full sized and could follow you around the house, just like another person could. I told them that I preferred the elfin rendition of persons on the other line, indeed would have found the life-sized ones disconcerting.

"It's pretty cool though, Auntie Grace," Faraoun told me enthusiastically. "It's kinda like a ghost, a friendly one. One you can talk to."

"Well, these life-sized realistic ghosts might be nice for a conversation," I told him, "But they would be singularly useless when it came to opening a jar for me or pulling a weed."

"Right," my nephew said, crestfallen, and we both knew what each other was thinking.

I slept better than I had right after Rob had died, for I was more used to the solitary life. But I missed my conversations with Botboy terribly and found myself talking to him in my head all the time. But I never could get him to answer.

I forced myself to keep cooking and eating, but it was a chore. I gardened still, but every sunset, every breathtaking view of the volcanoes on the horizon, every little green frog that made me smile with happiness only made the fact that I had no one to share them with more acute. Botboy had often said that humans were compelled by their socially constructed brains to want to *share* happiness. Humans, he said, felt that all happiness if enjoyed alone was diminished.

I can report that it was so.

I turned down many opportunities to speak about social robots. I knew I should do it, for the whole concept had not died just because my own robot had, but I simply could not bear it. Perhaps some day.

The security that had been in place, the silent guards who had patrolled the perimeter of our property, melted away. What was there to protect, now that the multi-million dollar robot had been destroyed? Mr Chukkerpuppy and his associates were behind bars.

And we were alone.

One Saturday when the boys were still in Sirhan's lab, O'Malley called me. He had actually sent me some flowers, green cymbidium orchids, my favorites and I had not yet thanked him.

"I hope you won't tell me I can't come, Grace," his warm voice said over the phone, a twelve-inch-high holographic image of him on my desk. I couldn't resist, I touched his hair, but of course it dissolved beneath my fingers. Unfortunately the hologram phone meant that the other person, if similarly equipped had a holographic image of you as well, but I preferred not to think about that or I might never have answered the phone. Today I had only my faded garden jeans on and my hair pulled back unattractively in a knot. O'Malley spoke as if he did not know these things.

"I want to come out next Saturday. I'd like to see Faraoun and Sirhan as well. I have a new project I'd like to discuss with them. And I'd like to see you."

His words hung in the air. I knew, could tell, that he was telling me that his interest in me was as a person, a woman, but I couldn't respond. I just couldn't. Rob was knit into every cell of my body. Would that ever change? I hoped not. Did it mean I would never have another relationship with a man? I didn't know and I couldn't think about it. As it was I felt twice widowed. But that seemed too silly to utter.

I told him he was welcome to come and pretended that I had to go take care of an urgent gardening task. He laughed, but did not seem to take offense.

Despite my multitudinous disclaimers, I found myself cleaning the house that week and even taking out colorful old clothes, once worn to make myself feel pretty.

Good grief, Grace, I thought. *Pretty? That's a laugh.*

The boys nodded when I told them O'Malley was coming out the following week, but they didn't seem that excited.

"We have to finish our project Friday night," Sirhan said, "so I doubt we'll be there when he gets here."

I looked at my nephew suspiciously. Project indeed. Were they imagining I wanted time alone with O'Malley?

Friday was a cool autumnal day, the morning was rainy, but the clouds burned off by early evening and the autumnal light on the maples which were now turning a deep scarlet and pink on the outside limbs, while remaining fresh green on those inside, caused my heart to ache with beauty. The geese flying south made me aware of the bite in the air. The three quail rushing out of the brush as I walked to check the mail made me laugh. Life.

It still surged through my no-longer-young veins and lifted my spirits.

I wanted a traditional dinner with roasted rosemary chicken, mashed potatoes, gravy, and biscuits. I made a salad out of the greens Botboy had planted in neat lines beneath the row cover, plating first a row, then waiting two weeks to plant another so that I would always have something fresh as the season wore on.

Gravel crunched in the driveway. That would be O'Malley. I was fussing with the salad and glanced out the window. He was driving an electric van and to my surprise had Faraoun and Sirhan with him. They must have met him at the airport without telling me. But where was their beat-up old car, the gasoline powered one that was the despair of Botboy's ecological conscience? The doors slammed, I could hear the van's side door slide open.

"It's unlocked," I called out, curiously reluctant to go to the door and seem so eager.

I heard the clump of men's feet as they entered the hall and took off their coats. But I heard something else was well. They must have brought someone else along. I felt a twinge of annoyance.

But annoyance turned to astonishment as they came around the corner, followed by none other than a robot. That robot looked a lot like Botboy, though his metal skin did not gleam as much.

I wiped my hands on a tea towel. Surely they would not expect me to accept a substitute for my very best friend? *Accept no substitutes* I told myself wryly.

O'Malley came forward, the boys hung back but their eyes gleamed suspiciously with suppressed what? Laughter? Excitement? O'Malley planted a kiss on my cheek and took my hand, drawing me forward to introduce this new robot.

"What is your name, Robot?" I asked politely.

The robot spun on his central axis, the way Botboy used to do when overjoyed, "She. Do you not recognize me? It is I. Botboy, your robot."

Despite the embedded cardiac bots my heart gave a lurch, "You cannot be my Botboy, he was destroyed in the mountain caves."

"So I have heard," this Botboy said, "But nonetheless, I am Botboy."

I looked helplessly at my nephew, at Sirhan, who now stepped forward

"Mrs Feenaughty. I have done a terrible thing. A wonderful thing. I was afraid of the military plan for Botboy. I felt he was never intended to come back from that mission and time proved me correct. Before he left I...I downloaded him. I cloned his 'brain.' This *is* Botboy. At least Botboy up to the time we left for Washington DC. He now has no memory of events after that."

Tears sprang to my eyes, "You downloaded him? My Botboy? You re-built him? How?"

Sirhan reached inside his jacket pocket and brought out the small bundle of Rob's notebooks. "The information you entrusted me with was our guide. Agent O'Malley has been helping us and in return will receive the notebooks so that other social robots can be built. Dr Feenaughty foresaw that we might need to...duplicate his efforts."

"Botboy." I was speechless, breathless. My robot rolled forward and took my hands into his white gloved ones.

"She." He said.

"But...but..." I looked past him to the boys again, "Isn't that replication? Isn't that forbidden?"

"I have broken every copyright, every ethical and every military law imaginable," Sirhan said "but I did it to keep Botboy safe. And Dr Feenaughty's dream of social robots alive."

"Botboy, how were *you* able to engage in forbidden behavior?" I asked wonderingly.

Botboy would have blushed if he could have, I am certain of it. "It *is* replication, She, and replication *is* forbidden. Let's change the subject."

We all laughed and with it the sorrow and the tension of those past months disappeared as if smoke up the chimney.

"Botboy you are being evasive," I teased.

"She, I can only do what is forbidden if that which I do is higher in the Rules Hierarchy than the forbidden thing. You see. replication is lower than the One Rule Founder told me never to disobey."

"What was that?" I asked, thinking of the ten commandments of my childhood catechism.

Take care of Grace Founder told me, *Take care of Grace.*

Botboy served our dinner as of old, the boys and O'Malley had apparently filled him in on the escapades of which he had no "memory" for he asked no questions. Instead he buzzed about the kitchen, perfectly at home, as if he had been gone an hour or two, correcting the salt I had added to the gravy, folding the clean napkins and placing them on the Aga to dry, wiping specks of dust from the windowsill. My heart sang with joy.

That night, after we five had talked our fill and eaten our fill and Botboy had opened the windows to let in the fragrant night air, I said good night to O'Malley and lay in my bed.

Sweet sleep began to descend as I listened to Botboy putter around in the laundry room. As I grew drowsier and drowsier I couldn't help my thoughts that circled in wonder at the mystery of the existence of "self" that went to the heart of "being." Did the fact that Botboy now had no memory of what we had been through, of things that "he" had done, alter the fact that he was Botboy? Was his new body only an empty shell, an *exoskeleton total*, a thing to be discarded at will or whenever necessity required? If I—or any human—were to shed our bodies, but download the contents of our brains, would we awaken as if from a long sleep with our immortal souls intact?

And as I closed my eyes an image rose up before me. It was of the monument erected to my robot, to his bravery, by our President. It was a simple monument. Into the gigantic grey stone that had been thrown by the blast against the mouth of the cave one word was carved in deep austere letters. One word for thousands of generations to read and ponder, for legends to embellish or even extinguish, for interplanetary travelers to discover: A monument in stone where the crushed body of my robot lay entombed, perhaps awaiting the resurrection of a future day. Guarding the silence, marking the sacrifice, that deeply carved word in letters three feet high said simply:

BOTBOY

Author's afterword

There I was, minding my own business, head in the clouds of medicine and science, as usual, driving to work. I clicked on a podcast of my favorite show: NPR's *Science Friday* with Ira Flatow. It was 2008 and my fifty-ninth birthday.

What would it be like, Ira asked, if you were to walk into your home and a robot was cooking dinner?

What would it be like? It would be great!

Would it be strange, Ira went on, if it asked you about your day? Or reminded you to call your mother? Or nagged you about watching too many football games?

Hmmm... not likely with the football games, but how would I feel? Strange? Delighted? Scared? Put off? Fascinated?

Fascinated, surely.

Ira spoke about the possibility of a "social" robot, one whose interactions with humans were based on conversation, perception of needs, feelings and emotion. Such a robot would have personality and would elicit a sense of connectedness from its humans. My physician's imagination raced forward with the possibilities such an advanced robot would offer for improved medical care, more accurate diagnosis, mind/body healing. My child's imagination (the one whose voice lives in all of us if we can only be still and stop editing for a moment) raced forward with other uses for such a "machine:" Alleviating loneliness, providing companionship, rescuing from danger.

And thus was Botboy born.

A social robot like Botboy, of immense complexity and hope, is yet technically beyond us. But that is today. Somewhere a youth has been born—or not quite yet been born—whose imagination and technical skills will give us breakthroughs to make social robotics just another branch of the natural and physical sciences. This youth will enthusiastically face the challenge of recreating what we take for granted: Hands that grasp and move fluidly and in concert with our eyes, eyes that scan the environment and retrieve information for the brain to process and integrate with unimaginable speed, brains that "think their own thoughts" at warp speeds and then both project and modulate those thoughts whenever around others. This robotic brain will have to read the emotion off of human faces, to hear, sense or smell deception.

The simple story of Botboy and Grace and Rob raises many questions, chief among them: What is the nature of consciousness? Can a robot who masters the rational use of pronouns such as "I" and "you" and (importantly) "we," be said to have consciousness? If it is able to pass the Turing test,[1] how is interacting with it different from interacting with another human? Is perfect emulation of us the same thing as being us? Or better? Or worse? Can a machine which lives to benefit me, which understands my needs, can respond to my questions, anticipate my desires and care for me without regard to it's own welfare, be said to "love" me? What is the nature of love? If a robot can tell a joke, pulling irony and incongruence from its database of

---

[1]The Turing test is a proposal for a test of a machine's ability to demonstrate intelligence. It proceeds as follows: A human judge engages in a natural language conversation with one human and one machine, each of which tries to appear human. All participants are placed in isolated locations. If the judge cannot reliably tell the machine from the human, the machine is said to have passed the test. (*Wikipedia*; Yang 8/2009)

human interaction, does it have a sense of humor? Clearly a machine can be programmed to learn and to teach. Is it then our mentor? Are we the teacher or, past some level of complexity, the taught?

And what does it mean to be "alive?" If my central nervous system, my brain, can be "downloaded" into another body, whereby my experiences and thoughts are passed on, do "I" live forever? Or can there be more than one of me if my brain is downloaded more than once into several different bodies? Are bodies even necessary at all for "us" to exist?

How important to being "alive" is replication? If robots can assemble other robots as RNA assembles proteins, is this the gift of life?

Oh yes. And what about death? Botboy makes us ask ourselves what it means to die.

These are the big questions Botboy raises, but there are more. Will an immensely complex social robot who "understands" how we think, understands our psychology, be controllable by humans or will it control us? Will humans be able to resist using these robots against other humans? Will they be agents of improving life on our planet or agents of destruction, conquest and warfare? Social robots are likely to be very expensive, does that mean that only certain economic classes of humans will have them? If their companionship is so wonderful, and they are so indestructible, why bother to relate to real people who will, after all, abandon us when they die?

While researching Botboy I came across several fascinating issues in robotics of which I had been unaware. Chief among these was the concept of the Uncanny Valley, an observation made by Masahiro Mori whereby we humans respond with attraction to robots the more they resemble us but only to a point. Once they are physically difficult to distinguish from us, they become repulsive and "uncanny." The valley is the dip in positive response seen on a graph as we near exact physical emulation.[2] I did not create Botboy as I did, with metal parts and a non-human face, for this reason, for Botboy sprang uncensored from my unconscious in the form for which I personally felt the most affection, but I was fascinated to understand why my brain may have done so.

For the interested reader of my little story of a robot and his humans there is literally a tsunami of information about robotics, artificial intelligence and virtual reality out there. How interested are you? Do you want to know about re-creating input to the first cranial nerve of the brain, the olfactory nerve?[3] [4] Or how about the role of memory, scent and olfaction?[5] Are you interested in the current thinking about social robots?[6] Perhaps you wish you had another life to give to science? If you are—or could be again—twenty years old, where in the world would you go to study robotics? The University of Tokyo? Waseda University? Vanderbilt? MIT? Carnegie Mellon? Or would you sequester yourself in your parents' basement and play play play with virtual reality games and avatars, letting your hair grow long and your school work languish?

---

[2]Mori, M. 1970 The Uncanny Valley (translated by K.F. MacDorman & T. Minato), *Energy*, 7 (4), pp 33-35

[3]Yanagida Y, Kawato S et al. Projection-Based Olfactory Display with Nose Tracking, *IEEE Virtual Reality Conference* 2004 (VR 2004)

[4]Cosimo D, Giovanni I, Giulio R An application of mobile robotics for olfactory monitoring of hazardous industrial sites, *Industrial Robot: An International Journal* 2009. 36 (1) pp 51-59

[5]http://www.macalester.edu/psychology/whathap/ubnrp/smell/memory.html A good, brief summary of the biology of smell and memory

[6] Fong T, Nourbakhsh I, Dautenhahn K A Survey of Socially Interactive Robots, *Robotics and Autonomous Systems 42 (2003) 143-166*

I am not twenty nor ever likely to be again, and my brain is not wired for entrance to MIT. Rather I am a physician who lives in the countryside, who collects plants and cares for children. When I sit in my chair by the fire my thoughts are unfettered; I am able to breathe life into Botboy with the only tool I really possess: Uncensored imagination.

Imagination lies, after all, at the core of scientific endeavor, its very core. For my scientific readers, let this story of imagination pull your more rational work along in its wake. Do not be surprised that a Perfectly Scientific Press would publish a story that has not yet, nor may ever, come to pass exactly in the way that I have imagined it. Consider fiction the wind beneath the wings of non-fiction, the wind beneath our wings.

Botboy, where are you? Humanity awaits.